CELL BIOLOGY RESEARCH PROGRESS

PHOTONIC HEMOSTASIS

PHYSIOLOGY OF LIGHT SIGNALS IN THE NEUTROPHIL

CELL BIOLOGY RESEARCH PROGRESS

Additional books in this series can be found on Nova's website
under the Series tab.

Additional e-books in this series can be found on Nova's website
under the e-book tab.

CELL BIOLOGY RESEARCH PROGRESS

PHOTONIC HEMOSTASIS

PHYSIOLOGY OF LIGHT SIGNALS IN THE NEUTROPHIL

THOMAS STIEF

New York

For permission to use material from this book please contact us:
Telephone 631-231-7269; Fax 631-231-8175
Web Site: http://www.novapublishers.com

Library of Congress Cataloging-in-Publication Data

ISBN: 978-1-62618-788-7

LCCN: 2013938079

Published by Nova Science Publishers, Inc. † New York

CONTENTS

PREFACE

NEUTROPHILS' PHOTONS AND OPSINS

Neutrophil granulocytes are the primary defence cells of blood against bacteria, fungi, parasites, or thrombi [1]. Their main weapons and signals are the reactive oxygen species (ROS) hydrogen peroxide (H_2O_2), the mother ROS, and hydroxyl-radical (•OH), non-radicalic excited singlet oxygen ($^1O_2^*$) the two daughter ROS. $^1O_2^*$ reacts with a C=C alkene group to form an excited carbonyl ($C{=}O^*$) that releases a photon with a wave length roughly in the violet spectrum (around 400 nm) (Table 1, Fig. 1) [2]. The assembly of their active NADPH-oxidase (the H_2O_2 generator), the few specific triggers and many specific or unspecific primers are of great physiological and patho-physiological importance in inflammation and in hemostasis. The neutrophils generate different types of photons and they can "see" them. 300-400 nm photons are the main signals, the photons of lowest wave length seem to especially alert the neutrophils in emergency. The neutrophils can also respond to photons generated by other cells in the blood stream, such as the endothelium [3-5] or thrombocytes [6-8].

What are the "eyes" of human cells [9]? The 11-*cis*-(all *trans*-)retinal-opsins or the flavo-cryptochromes (*cry1, cry2;* also present in the inner human retina) can sense photons [10]. An opsin is a seven-helix-transmembrane receptor that is coupled to a G-protein (GPCR) [11] (Fig. 2). There are many important receptors in human physiology that are members of this family (e.g. the epinephrine, the bradykinin, the prostaglandin, or the angiotensin-2 receptor [13,14]). The medical importance of these 7-helix-receptors is immense [15]. Usually they are coupled to G-proteins, but there are other intracellular partners, and there is a complex interplay of positive and negative regulations [15]. 1% of our genome are GPCRs (about 1000 different receptor proteins). Humans feel the world through GPCRs [15]. 7 membrane-spanning domains (TM1-TM7) are connected by 3 extracellular (e1, e2, e3) and 3 intracellular (i1, i2, i3) loops (Fig. 2) [15]. The binding site of a molecular GPCR ligand is usually localized in e3 [13].

The receptor detects the presence of the "first messenger" (e.g. a hormone) outside the cell. Then the associated trimeric G-protein dissociates into the Gα subunit (linked to GTP) and the Gßγ subunit. One of the free G-subunits activates an enzyme or an ion channel. Second messengers are formed and the signal is transmitted intracellularly [15]. The 4 phospholipase C (PLC) ß-isoforms are regulated by G-proteins. PLC hydrolyzes the phospholipid phosphatidyl-inositol to get free inositol 1,4,5-triphosphate (the Ca^{2+} efflux controller) and diacylglycerol (DAG; the protein kinase C activator). ß-arrestins might function as alternative transducers of GPCR signals [15].

There are one or more N-glycosylation sites in the N-terminus of the GPCRs. Two cysteines form a disulphide bridge between external loop 1 and external loop 2. Another cysteine in the intracellular C-terminus is the site for palmitoylation, generating a 4th intracellular loop. There are about 20 different Gα proteins that either stimulate (Gαs) or inhibit (Gαi; typical for OPN5) the adenylate cyclase, or regulate phospholipase C – controlled Ca^{2+} fluxes (Gαq, Gα11, Gα14, Gα15), or via Rho-GTPases regulate cytoskeletal assembly priming the neutrophil for NADPH-oxidase activation (Gα12, Gα13). There are also at least 5 different Gß subunits and 12 Gγ subunits combining (randomly) to Gßγ dimers [15].

The 5 light sensing opsins are OPN1, OPN2, OPN3, OPN4, OPN5. OPN1 is the receptor of cones for red, green, or blue light with absorbance maxima for long wavelengths (LW; Xq28 chromosomal location), middle wavelengths (MW; Xq28), or short wavelengths (SW_1; 7q32.1), respectively [11].

These opsins for visible light to the human eye are also called phot-opsins. OPN2 (RHO) is rhodopsin, the receptor of rods for brightness of light (3q22.1). Figure 3 illustrates the respective absorbance maxima of OPN1 or OPN2 receptors. Table 1 characterizes the for the human eye visible electromagnetic wavelengths of 380-780 nm. Violet photons are about twice as energetic as red ones.

In the "3 sites rule" there are only 3 amino acids at sites 180, 277, and 285 (most importantly the latter two) that determine if the opsin senses red or green light: if the amino acids at the 3 sites are AFA (alanine… phenylalanine…alanine) then photons are recognized as red, if the amino acids are SYT (serine…tyrosine…threonine) then photons are sensed as green; a blue shift in the λmax values occurs by histidine→tyrosine at site 197 and alanine→serine at site 308 in other mammals ("5 sites rule") [16]. Interestingly, the LW and MW receptors of mammals have a second minor λmax at about 380 nm in the UV spectrum, this second minor λmax might be especially pronounced in fish [16].

Table 1. Physical characteristics of light [12]

Colour	Electromagnetic wavelength [nm]	Frequency [THz]	Energy per photon [eV]
Violet	380-430	697-789	2.88-3.3
Blue	430-490	612-697	2.53-2.88
Green	490-570	526-612	2.17-2.53
Yellow	570-600	500-526	2.06-2.17
Orange	600-640	468-500	1.95-2.06
Red	640-780	384-468	1.6-1.95

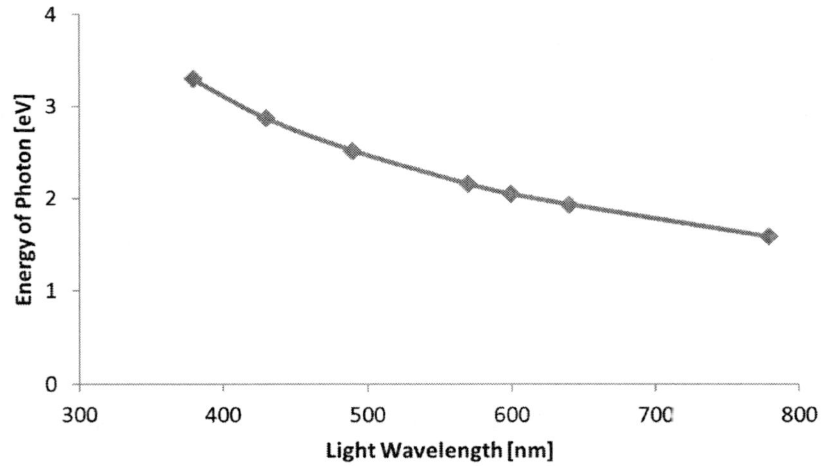

Figure 1. Photonic energy. The data of Table 1 are plotted. The x points are the inferior and superior range limits of violet, blue, yellow, orange, or red light. Photons in the ultraviolet (UV) spectrum at about 300 nm have an energy of about 4 eV.

OPN3 is encephalopsin (panopsin; 1q43) [11]. OPN4 is melanopsin (10q23.2) an opsin with bistable sequential photon absorption, especially found in the intrinsically photosensitive retinal ganglion cell (ipRGC); extrinsic = rods and cones, intrinsic = ipRGC; retinal melanopsin is 10.000--fold less concentrated than rhodopsin [17-21]. OPN5 is the receptor of ultraviolet light (SW_1) [22]. The usual direct photon receptor for light sensing opsins is 11-*cis*-retinal but there are opsins that work with all-*trans*-retinal, e.g. the photoisomerases that convert all-*trans*-retinal into the biologically more important 11-*cis*-retinal. Peropsin (RRH, OPSX; 4q25) [23], retinal G-protein coupled receptor (RGR; 10q23.1; RPE65 = retinal pigment epithelium-specific protein 65kDa; LCA2, RP20, rd12, RBP-binding membrane protein) are photoisomerases [11,24]. An important opsin without known light sensing is neuropsin (6p12.3), an opsin involved e.g. in neurologic functions of intelligence and memory [25].

11-*cis*-retinal upon capturing a photon changes its conformation (finally into all-*trans*-retinal), the opsin structure changes with the chromophore in the form of meta 2 (Fig. 4): helices 3 and 6 are moved outward, affecting G-protein (transducin) binding sites such as the (in Fig. 3 turquoise blue coloured) cytoplasmic loop between helices 5 and 6 [26]. Transducin is activated and the chromophore changes to all-trans-retinal (Figure 3).

The E113 stabilizer might prevent that a low-energetic photon activates the receptor of high-energetic photons. There might be a photons quencher such as tyrosine, comparable to a "lightning conductor" in a cluster of key residues [27] that prevents that a high-energetic photon activates the receptor of low-energetic photons. So, red photons would not activate the blue receptor and viceversa.

The coupled protein transducin (a G-protein) is activated, cGMP phosphodiesterase is stimulated, the intracellular cGMP concentration decreases, cGMP-gated sodium channels close, the visual photoreceptor cell is hyperpolarized, and second retinal neurons communicate the seen information to the brain [11,15,28]. Activated Gαs also stimulate "L-type" Ca^{2+} channels or activated Gαo subtypes inhibit high voltage "N-type" Ca^{2+} channels [15]. In melanopsin (OPN4) signalling a Gq11,14/PLC cascade is triggered that induces depolarizing cellular currents by activation of transient receptor potential (TRP) channels (TRP6 and TRP7 similar as in rhabdomeric opsins) that may be functionally redundant [10]. Compared with OPN1 both the activation and the inactivation of OPN4 is very slow. There are latencies of seconds from "light on" to firing and of tens of seconds from "light off" to stop firing. The light intensity determines the maximal and steady state cellular firing rates, OPN4 being more an irradiance detector than a vision receptor [10]. Lecithin:retinol acyl-transferase (LRAT) and especially RPE65 generate the necessary 11-cis-retinal; like rhabdomeric opsins OPN4 is bistable, i.e. OPN4 works both with 11-cis-retinal and all-trans-retinal [10].

In OPN1/11-*cis*-retinal transducin is activated at the meta 2 phase (λmax = 380 nm), afterwards there appears a meta 3 form (λmax = 450 nm) [28], suggesting that capturing UV/violet/blue photons could contribute to reisomerize all-*trans*-retinal to 11-*cis*-retinal. An UV-illuminated OPN5 receptor changes its λmax from 380 nm to 470 nm, a stable molecule in the dark (bistability) that reverts to λmax of 380 nm upon illumination with orange light [10], suggesting that capture of LW photons reisomerizes SW_2 receptors. The light transduction is down-regulated by phosphorylases following GPCR activation [29,30] (e.g. by GPRK1 =G protein – coupled receptor kinase 1; RHOK, RK, rhodopsin kinase), by arrestin-adaptin-clathrin-dynamin endocytosis of GPCRs, by changed lipid profile anchoring the membraneous receptor [31], by loss of N-linked glycans [32] and by receptor ubiquitination and destruction in the proteasome. Second messengers can also be neutralized by cAMP phosphodiesterases, phosphatidyl-inositol phosphatases, and DAG kinases [15].

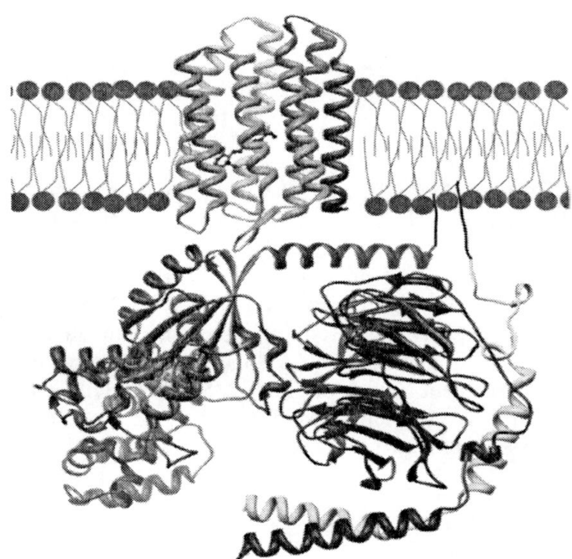

Figure 2. Rhodopsin in the plasma membrane and the G-protein inside the cell [12]. The 11-cis-retinal-opsin of rods is a 7-helix-receptor with the black coloured chromophore 11-cis-retinal Schiff-base-linked to lysine (K296) in the seventh helix (TM7) inside the plasma membrane. The stabilizer of the protonated positively charged Schiff-base is the negatively charged glutamic acid (E113). The G-protein transducin (Gt) is inside the cell below the opsin that is embedded in the lipid bilayer. Gtα (with a bound GDP) is colored in red, Gtß in blue, and the main membrane anchor Gtγ [15] in yellow. The C-terminus of the opsin is inside the cell, the N-terminus is outside the cell [15,37]. Opsins function by light absorption and G-protein coupled receptor (GPCR)-dimerization enhances signal transmission [14].

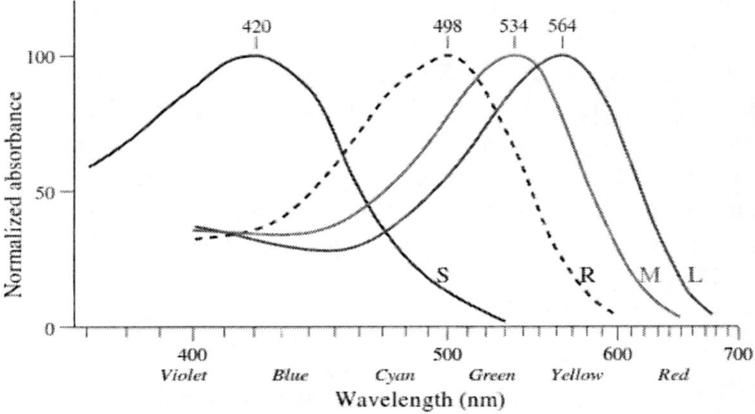

Figure 3. Absorption maxima of the 3 human phot-opsins and of rhodopsin [12]. The absorption maximum of the blue light receptor is at 419-420 nm (SW_1), that of the green light receptor is at 531-534 nm (MW), that of the red light receptor is at 558-564 nm (LW), and that of rhodopsin is at 496-498 nm, OPN4 has a λmax of about 479 nm, the λmax of OPN5 (SW_2) is 380 nm [22]. Sometimes, the chromophore-opsin is triggered outside its λmax. The brain sensation "red" is generated, if a photon of light of 650-700 nm wavelength triggers the OPN1-LW receptor. Interestingly, the physical red light wavelength is about 100 nm higher than the λmax of OPN-LW. The brain sensation “yellow” is generated by a combined activation of OPN1-LW and OPN1-MW by electromagnetic waves of about 585 nm. The brain is the “artist” that mixes all colors out of different excitations of its red-, green-, and blue- receptors. The human neutrophil possesses the two opsins that sense SW_1 and SW_2 photons.

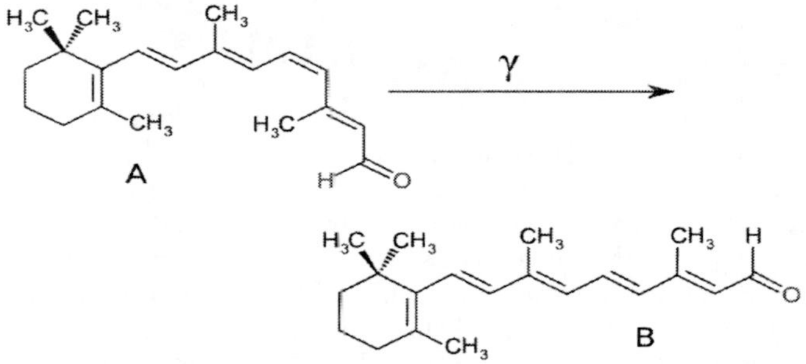

Figure 4. Conversion of 11-cis-retinal into all-trans-retinal [12]. 11-cis-retinal (A) captures at C-atom nr. 5 (C-atom nr. 1 being the highest oxidized C-atom, i.e. the one with the aldehyde group) one photon (γ), which changes the molecular structure into all-trans-retinal (B). A→B intermediates are photo (570 nm λmax), batho (543 nm), lumi (497 nm), meta 1 (480 nm), meta 2 (380 nm) that appear femtoseconds, picoseconds, nanoseconds, microseconds, milliseconds (respectively) after a photon reaction with 11-*cis*-retinal [15,28,33-37].

This book is on the regulation of the neutrophil's ROS generation by different photons, by singlet oxygen (the excited "pro-drug" of photons), by important proteins, or by drugs (e.g. modulators of the eicosanoid metabolism) that should not generate systemically circulating micro-thrombi.

The book chapters:

1. Light quants of low wave length prime blood neutrophils for ROS generation
2. Blood neutrophils see UV light: 340 nm ultraviolet A stimulates blood ROS generation nearly half as strong as 405 nm violet photons
3. Blood neutrophils alert each other by photons
4. Singlet oxygen ($^1O_2^*$) primes blood neutrophils to generate ROS
5. Singlet oxygen – oxidized human albumin stimulates blood ROS generation
6. Human IgG can modulate blood ROS generation
7. Naproxen is only a mild inhibitor of neutrophils
8. Therapeutic human antithrombin-3 inhibits blood ROS generation
9. Thrombin generation by naproxen

Releasable photon of excited non-radical singlet oxygen ($^1\Delta O_2^*$).

11-*cis*-retinal-opsin has just received a photon.

Priv.-Doz. Dr. med. Thomas Stief
Institute of Laboratory Medicine and Pathobiochemistry
University Hospital
D-35043 Marburg
Germany
Tel.: +49-6421-58 64471
Email: thstief@t-online.de

REFERENCES

[1] Stief TW. Neutrophil granulocytes in hemostasis. *Hemostasis Laboratory* 2008; 1: 269-89.

[2] Stief TW. Regulation of hemostasis by singlet oxygen ($^1\Delta O_2$). *Curr Vasc Pharmacol* 2004; 2: 357-62.

[3] Jansen F, Yang X, Franklin BS, Hoelscher M, Schmitz T, Bedorf J, Nickenig G, Werner N. High glucose condition increases NADPH oxidase activity in endothelial microparticles that promote vascular inflammation. Cardiovasc Res. 2013; 98: 94-106.

[4] Farjo KM, Farjo RA, Halsey S, Moiseyev G, Ma JX. Retinol-binding protein 4 induces inflammation in human endothelial cells by an NADPH oxidase- and nuclear factor kappa B-dependent and retinol-independent mechanism. *Mol Cell Biol*. 2012; 32: 5103-15.

[5] Schramm A, Matusik P, Osmenda G, Guzik TJ. Targeting NADPH oxidases in vascular pharmacology. *Vascul Pharmacol*. 2012; 56: 216-31.

[6] Ruf A, Schlenk RF, Maras A, Morgenstern E, Patscheke H. Contact-induced neutrophil activation by platelets in human cell suspensions and whole blood. *Blood*. 1992; 80: 1238-46.

[7] Amer J, Zelig O, Fibach E. Oxidative status of red blood cells, neutrophils, and platelets in paroxysmal nocturnal hemoglobinuria. *Exp Hematol*. 2008; 36: 369-77.

[8] Lopez JJ, Salido GM, Gómez-Arteta E, Rosado JA, Pariente JA. Thrombin induces apoptotic events through the generation of reactive oxygen species in human platelets. *J Thromb Haemost*. 2007; 5: 1283-91.

[9] Stief TW. The blood fibrinolysis / deep-sea analogy: a hypothesis on the cell signals singlet oxygen/photons as natural antithrombotics. *Thromb Res*. 2000; 99: 1-20.

[10] Sexton T, Buhr E, Van Gelder RN. Melanopsin and mechanisms of non-visual ocular photoreception. *J Biol Chem*. 2012; 287: 1649-56.

[11] Terakita A. The opsins. *Genome Biology* 2005; 6: 213.

[12] www.wikipedia.org

[13] Miura S, Imaizumi S, Saku K. Recent progress in molecular mechanisms of angiotensin II type 1 and 2 receptors. *Curr Pharm Des*. 2012; 19: 2981-7.

[14] Audet M, Bouvier M. Restructuring G-protein- coupled receptor activation. *Cell*. 2012; 151: 14-23.

[15] Luttrell LM. Reviews in molecular biology and biotechnology: transmembrane signaling by G protein-coupled receptors. *Mol Biotechnol*. 2008; 39: 239-64.

[16] Yokoyama S, Radlwimmer FB. The molecular genetics of red and green color vision in mammals. *Genetics* 1999; 153: 919-32.

[17] Jiang M, Pandey S, Fong HK. An opsin homologue in the retina and pigment epithelium. *Invest Ophthalmol Vis Sci* 1993; 34: 3669-78.

[18] Peirson S, Foster RG. Melanopsin: another way of signaling light. *Neuron*. 2006; 49: 331-9.

[19] Hughes S, Hankins MW, Foster RG, Peirson SN. Melanopsin phototransduction: slowly emerging from the dark. *Prog Brain Res*. 2012; 199: 19-40.

[20] Lucas RJ, Lall GS, Allen AE, Brown TM. How rod, cone, and melanopsin photoreceptors come together to enlighten the mammalian circadian clock. *Prog Brain Res*. 2012; 199: 1-18.

[21] Do MT, Yau KW. Intrinsically photosensitive retinal ganglion cells. *Physiol Rev*. 2010; 90: 1547-81.

[22] Kojima D, Mori S, Torii M, Wada A, Morishita R, Fukada Y. UV-sensitive photoreceptor protein OPN5 in humans and mice. *PlosOne* 2011; 6: e26388.doi.10.1371/journal.pone.0026388.

[23] Sun H, Gilbert DJ, Copeland NG, Jenkins NA, Nathans J. Peropsin, a novel visual pigment-like protein located in the apical microvilli of the retinal pigment epithelium. *Proc Natl Acad Sci USA*. 1997; 94: 9893-8.

[24] Pepe IM, Cugnoli C. Retinal photoisomerase: role in invertebrate visual cells. *J Photochem Photobiol B* 1992; 13: 5-17.

[25] Izumi A, Ijima Y, Noguchi H, Numakawa T, Okada T, Hon H, Kato T, Tatsumi M, Kosuga A, Kamijima K, Asada T, Arima K, Saitoh O, Shiosaka S, Kunigi H. Genetic variations of human neuropsin gene and psychiatric disorders: polymorphism screening and possible association with bipolar disorder and cognitive functions. *Neuropsychopharmacology* 2008; 33: 3237-45.

[26] Bhattacharya S, Hall SE, Vaidehi N. Agonist-induced conformational changes in bovine rhodopsin: insight into activation of G-protein-coupled receptors. *J Mol Biol*. 2008; 382: 539-55.

[27] Palczewski K, Kumasaka T, Hori T, Behnke CA, Motoshima H, Fox BA, Le Trong I, Teller DC, Okada T, Stenkamp RE, Yamamoto M, Miyano M. Crystal structure of rhodopsin: a G protein-coupled receptor. *Science* 2000; 289: 739-45.

[28] Hernandez-Rodriguez EW, Sanchez-Garcia E, Crespo-Otero R, Montero-Alejo AL, Montero LA, Thiel W. Understanding rhodopsin mutations linked to the retinitis pigmentosa disease: a QM/MM and DFT/MRCI study. *J Phys Chem B*. 2012; 116: 1060-76.

[29] Gehret AU, Hinkle PM. Importance of regions outside the cytoplasmic tail of G-protein-coupled receptors for phosphorylation and dephosphorylation. *Biochem J*. 2010; 428: 235-45.

[30] Butcher AJ, Kong KC, Prihandoko R, Tobin AB. Physiological role of G-protein coupled receptor phosphorylation. *Handb Exp Pharmacol*. 2012; 208: 79-94.

[31] Oates J, Watts A. Uncovering the intimate relationship between lipids, cholesterol and GPCR activation. *Curr Opin Struct Biol*. 2011; 21: 802-7.

[32] Kroeger H, Chiang WC, Lin JH. Endoplasmic reticulum-associated degradation (ERAD) of misfolded glycoproteins and mutant P23H rhodopsin in photoreceptor cells. *Adv Exp Med Biol*. 2012; 723: 559-65.

[33] Liu X, Garriga P, Khorana HG. Structure and function in rhodopsin: correct folding and misfolding in two point mutants in the intradiscal domain of rhodopsin identified in retinitis pigmentosa. *Proc Natl Acad Sci USA*. 1996; 93: 4554-9.

[34] Okada T, Palczewski K. Crystal structure of rhodopsin: implications for vision and beyond. *Curr Opin Struct Biol*. 2001; 11: 420-6.

[35] Smith SO. Insights into the activation mechanism of the visual receptor rhodopsin. *Biochem Soc Trans*. 2012; 40: 389-93.

[36] Zhou XE, Melcher K, Xu HE. Structure and activation of rhodopsin. *Acta Pharmacol Sin*. 2012; 33: 291-9.

[37] Palczewski K. Chemistry and biology of vision. *J Biol Chem*. 2012; 287: 1612-9.

Chapter 1

LIGHT QUANTS OF LOW WAVE LENGTH PRIME BLOOD NEUTROPHILS FOR ROS GENERATION

ABSTRACT

Background: Neutrophils are the effector cells of fibrinolysis (tertiary hemostasis). They generate the excited (*) oxygen species singlet oxygen ($^1\Delta O_2^*$), a selectively destructive and signaling reactive oxygen species (ROS) that emits light quants. Since the neutrophils might not only generate but also perceive the inter-cellular communication signal light quants, photons of different wave lengths were analyzed for a possible pro-ROS generation action on blood neutrophils.

Material and Methods: 10 µl fresh citrated normal blood were added in 4-fold to transparent polystyrene U-microwells (Brand®781600) prefilled with 150 µl Hanks′ Balanced Salt Solution. The blood samples were excited 0-5-fold by light of the wave lengths 405 nm (violet), 450 nm (blue), 492 nm (green), or 620 nm (orange) using a microtiter plate photometer (PHOmo). 100 µl reaction mixture (6.3 µl blood) of the transparent wells were withdrawn and transferred into black polystyrene F-microwells (Brand®781608). 0.8 µg/ml (final) zymosan A (ZyA) or no specific ROS generation trigger was added. After addition of 0.4 mM (final) luminol the luminescence was measured. After 100 min (37°C) half of the wells without ROS generation trigger were supplemented with 3 µg/ml ZyA (final) and the luminescence was determined.

Results: 0.8 µg/ml ZyA-incubated normal blood that had previously been 405 nm excited 0, 1, 2, 3, 4, 5-fold had a ROS generation at 67 min of 86, 236, 297, 468, 547, 983 RLU/s, respectively. 3-fold excitation with a wavelength of 450-620 nm could only double ZyA-triggered blood

ROS generation, 5-fold excitation resulted in lower ROS generations than unexcitede blood samples. If excited blood was pre-incubated for 100 min (37°C) in absence of a ROS generation trigger, and then incubated in presence of 3 µg/ml ZyA, very high ROS generations appeared: 100 min of incubation resulted in maxima of 2961 to 12602 RLU/s for 0 to 5-fold 405 nm excited blood. Without the ZyA-trigger the normal blood ROS generation was always 0 RLU/s, independent of wavelength, excitation time, pre-incubation time, or incubation time.

Discussion: Photons of about 400-425 nm wavelength (violet/blue) prime neutrophils for ROS generation. This is of great importance in the physiological and pathophysiological understanding of inflammation. Physiological inflammation as e.g. in thrombosis could be stimulated by intra-venous "injection" of violet/blue light, in pathological inflammation violet/blue photon generation or action must be inhibited.

Keywords: Neutrophils, light quants, photons, 405 nm, violet, primer, stimulator, ROS, fibrinolysis

INTRODUCTION

Pathological activities of systemic thrombin generate increased concentrations of circulating micro-thrombi that are deposited in the vital organs causing renal, pulmonary, cerebral, hepatic, or myocardial dysfunction, unless these micro-thrombi are not attacked by tertiary hemostasis (fibrinolysis) [1,2]. Fibrinolysis has a plasmatic part [3] and a cellular one, the activated neutrophils being the main cells of cellular fibrinolysis [4] with urokinase and plasmin [5,6] as the central serine proteases in both fibrinolysis parts.

Sufficient fibrinolysis requires a non-frustrating action of the neutrophils [7]. Reactive oxygen species (ROS) generated by membranous NADPH-oxidase assembly together with urokinase, bound to the urokinase receptor that is coupled to the "innate non-self" receptor CD11b/18 [8-11] are the main fibrinolytic weapons of neutrophils [12-16]. Reactive oxygen species are here understood as derivatives of molecular oxygen (O_2) that react with target molecules. The main primary ROS is H_2O_2, the main secondary ROS are hydroxyl radical •OH and excited singlet oxygen ($^1\Delta O_2^*$). An intermediate oxygen derivative in the genesis of ROS is superoxide anion radical ($O_2^{\cdot -}$). The ROS are physiologically generated by normal activities of NADPH-oxidases (Nox1 to Nox5; the neutrophils′ Nox2 acts together with the neutrophils´ myeloperoxidase) but they can also be formed by a "leaky" action of cytochrome P450 (CYP450) or of mitochondrial respiration [12]. Mechanism based inhibited CYP450 [13,14] or uncoupled NO• synthase might also participate in the pathogenesis of ROS [15,16]. In low concentrations the ROS are cell signals, in high concentrations they are toxic to invading germs or own tissue [4,17-19].

Thrombi activate blood neutrophils to generate ROS [20] and thus call for augmentation of physiologic inflammation. ROS induce slow rolling and the firm adhesion of neutrophils to the endothelium [21-25] or the invasion into inflamed tissue [26]. The main ROS in physiologically enhanced neutrophils/endothelium interaction seems to be singlet oxygen ($^1\Delta O_2^*$) [27-29]. $^1\Delta O_2^*$ shifts hemostasis into an anti-thrombotic state inhibiting coagulation and stimulating fibrinolysis [30-43]. Since the neutrophils might not only generate but also perceive the inter-cellular communication signal light quants, photons of different wave lengths were here analyzed for a possible blood ROS generation priming in the new blood ROS generation assay (BRGA) [44,45].

Material and Methods

10 μl fresh (aged less than 1h) normal blood anticoagulated with 11 mM Na_3 - citrate, pH 7.4, in polypropylene tube (time to expiry: about 1 year left; Sarstedt, Nümbrecht, Germany) were added in 4-fold using an Eppendorf-multipette® after written informed consent to transparent high quality polystyrene U-microwells (Brand, Wertheim, Germany; article nr. 781600) prefilled with 150 μl Hanks´ Balanced Salt Solution (HBSS; modified without phenol red; SAFC Biosciences-Sigma, Deisenhofen, Germany; article nr. 55037C-1000ML).

The blood samples were excited 0-5-fold by a microtiter plate photometer (PHOmo; anthos, Krefeld, Germany), using light of the wave lengths 405 nm (violet), 450 nm (blue), 492 nm (green), or 620 nm (orange). 100 μl reaction mixture (6.3 μl blood) of the transparent wells were withdrawn and transferred into black high quality polystyrene F-microwells (Brand; article nr. 781608). 5 μl 18 μg/ml zymosan A (ZyA, 0.8 μg/ml final; Sigma, Deisenhofen, Germany; article nr. Z-4250-1G, lot nr. 27H0495) in 0.9% NaCl or no specific ROS generation trigger was added. 10 μl 5 mM luminol sodium salt (0.4 mM final; Sigma) in 0.9% NaCl were added and the luminescence in relative light units per second (RLU/s) was measured by a microtiter plate luminometer (LUmo; anthos) after 0-216 min (37°C). After 100 min (37°C) pre-incubation half of the wells without ROS generation trigger were supplemented with 10 µl 36 µg/ml ZyA (3 µg/ml final) in 0.9% NaCl and the luminescence was determined after 0-116 min incubation.

The mean values of the 4-fold determinations were calculated. The intra-assay coefficients of variation were less than 10%.

HBSS consisted of 185.4 mg/l $CaCl_2 \cdot 2\,H_2O$, 200 mg/l $MgSO_4 \cdot 7\,H_2O$, 400 mg/l KCl, 60 mg/l KH_2PO_4, 350 mg/l $NaHCO_3$, 8000 mg/l NaCl, 90 mg/l Na_2HPO_4, 1000 mg/l glucose, pH 7.0-7.4. Expressed in molarity, the concentrations of the HBSS components are: 1.3 mM Ca^{2+}, 0.8 mM Mg^{2+}, 5.8 mM K^+, 143 mM Na^+, 144 mM Cl^-, 1.6 mM SO_4^{2-}, 0.4 mM $H_2PO_4^-$, 0.6 mM HPO_4^{2-}, 4.2 mM HCO_3^-, 5.6 mM glucose.

The BRGA is an ideal tool to study ROS generation [44,45] because it

1. imitates the physiologic ROS generation response to blood pathogens
2. imitates the pathophysiologic blood ROS generation in autoimmunity
3. detects globally the most important ROS of blood, i.e. $H_2O_2 + \cdot OH + {}^1O_2$
4. uses the stable pathophysiologic trigger hydrophilic ZyA
5. is sensitive enough to detect ROS upon blood stimulation by only about 1 µg/ml ZyA (the approx. conc. of initial fungal sepsis [46-48])
6. does not interfere with cell function (using untoxic luminol concentrations)
7. is cheap: without taxes one LUmo (microtiter plate luminometer) costs less than 4500 €
8. uses a close-to-physiologic buffered salt solution [49]
9. is easy to standardize
10. works with whole blood
11. is quantitative and can therefore compare the results with the normal range (MV±1SD)
12. is suitable for routine measurements of hundreds of samples within minutes.

RESULTS

0.8 μg/ml ZyA-incubated normal blood without exposure to 405 nm photons generated maximally 354 RLU/s at 114 min (37°C). At 216 min only about 50% of the maximal blood ROS generation was achieved (Figure 1). The ROS values found at 132 min were 325, 576, 658, 1420, 1627, 1856 RLU/s for blood samples 405 nm excited 0-fold, 1-fold, 2-fold, 3-fold, 4-fold, 5-fold, respectively. Interestingly, the at least 2-fold excited samples had no decrease phase of blood ROS generation. Without ZyA addition the blood ROS generation was always 0 RLU/s, independent of wavelength, excitation time, or incubation time.

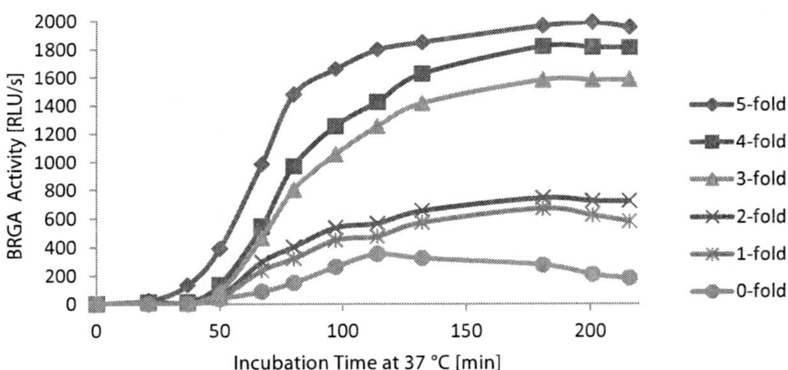

Figure 1. Kinetic of blood ROS generation in 405 nm excited blood. 10 μl fresh citrated normal blood were added in 4-fold to transparent polystyrene U-microwells (Brand®781600) prefilled with 150 μl Hanks′ Balanced Salt Solution (HBSS). The microtiter plates were excited at 405 nm 0-fold, 1-fold, 2-fold, 3-fold, 4-fold, or 5-fold by a microtiter plate photometer (PHOmo). 100 μl reaction mixture of the transparent wells were withdrawn and transferred into black polystyrene F-microwells (Brand®781608). 0.8 μg/ml zymosan A (final) and 0.4 mM luminol (final) were added and the luminescence (RLU/s) were measured by a microtiter plate luminometer (LUmo) after 0-216 min (37°C).

The more the blood samples were excited the more ROS were generated (Figure 2): 5-fold 405 nm excitation resulted in about 10-fold increased ZyA-triggered blood ROS generation. The blood ROS generations at 67 min were 86, 236, 297, 468, 547, 983 RLU/s for normal blood 405 nm excited 0-fold, 1-fold, 2-fold, 3-fold, 4-fold, 5-fold, respectively.

Wavelengths ≥ 450 nm generated much less ROS upon ZyA-stimulation than 405 nm (Figure 3). 3-fold excitation with a wavelength of 450 nm (blue), 492 nm (green), or 620 nm (orange) could double ZyA-triggered blood ROS generation. 5-fold excitation resulted in lower blood ROS generation than non-excited blood samples.

If excited blood was pre-incubated for 100 min (37°C) without the presence of a ROS generation trigger, and then incubated in presence of a specific trigger (3 μg/ml ZyA), considerably higher ROS generations could be measured: 100 min of incubation resulted in maxima of 2961, 7273, 8142, 9939, 10965, or 12602 RLU/s for blood 405 nm excited 0-fold, 1-fold, 2-fold, 3-fold, 4-fold, or 5-fold, respectively (Figure 4).

In another setting 405 nm excitation increased ZyA-triggered blood ROS generation, too. 5-fold excitation resulted in about 4-fold increased generation of blood ROS (Figure 5).

Wavelengths ≥ 450 nm only slightly primed neutrophils for ZyA-triggered ROS generation. 3-fold excitation with such wavelengths could only double ZyA-triggered blood ROS generation (Figure 6). Without the ZyA-trigger the normal blood ROS generation was always 0 RLU/s, independent of wavelength, excitation time, pre-incubation time, or incubation time.

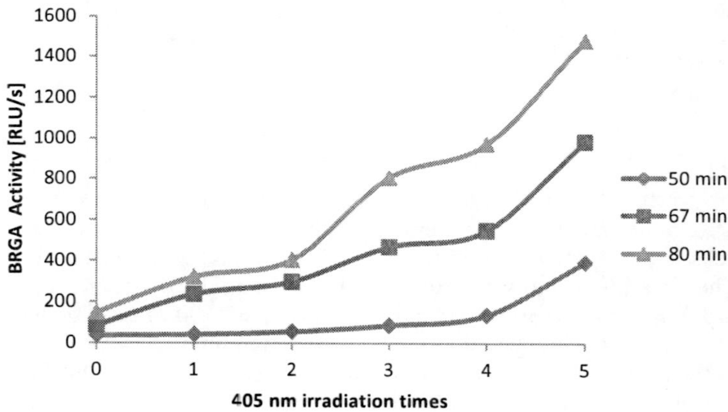

Figure 2. BRGA activity in dependence of 405 nm excitation times. The results of figure 1 were reproduced with 50 min, 67 min, or 80 min incubation time. 5-fold 405 nm excitation by the PHOmo microplate photometer resulted in about 10-fold increased blood ROS generation assay (BRGA) activity, determined in relative light units per second (RLU/s).

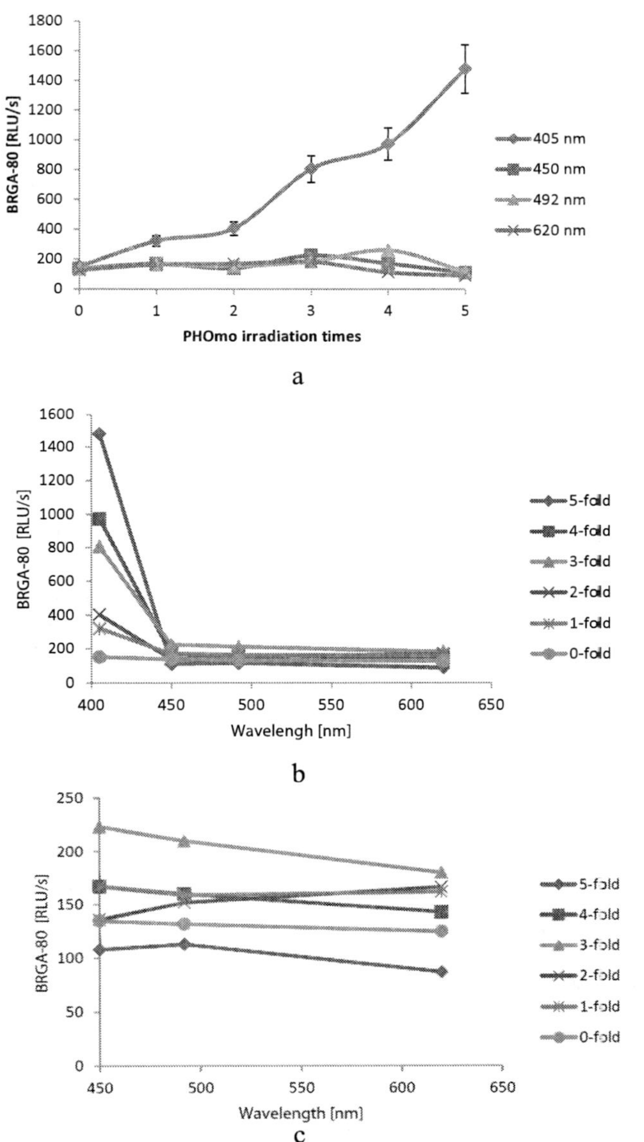

Figure 3. Wavelength dependence of the increase in blood ROS generation. The experiment of figure 1 with 80 min incubation time (BRGA-80) was performed with irraditions at the wavelengths 405 nm (violet), 450 nm (blue), 492 nm (green), or 620 nm (orange). 3-fold excitation at $\geq$ 450 nm could double ZyA-stimulated blood ROS generation. 5-fold excitation resulted in lower ZyA-stimulated blood ROS generation than unradiated blood samples.

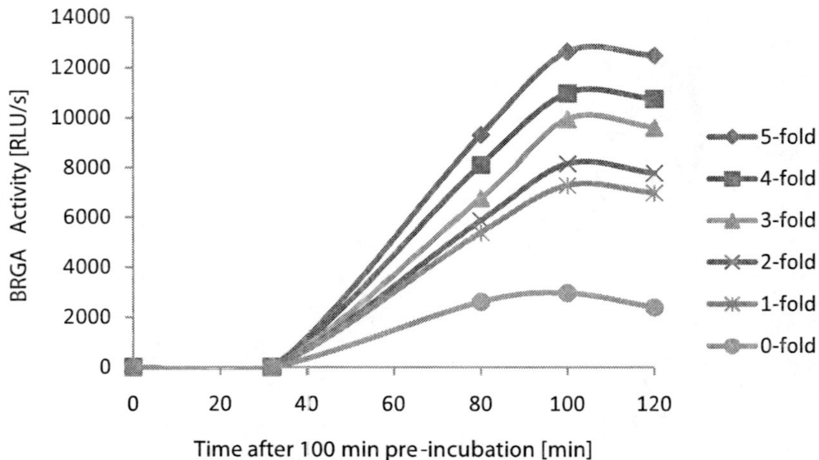

Figure 4. Kinetic of blood ROS generation in 405 nm excited blood after 100 min pre-incubation. After 100 min (37°C) pre-incubation as described in figure 1 however without addition of a ROS generation trigger, 3 μg/ml ZyA (final) were added and the luminescence was determined.

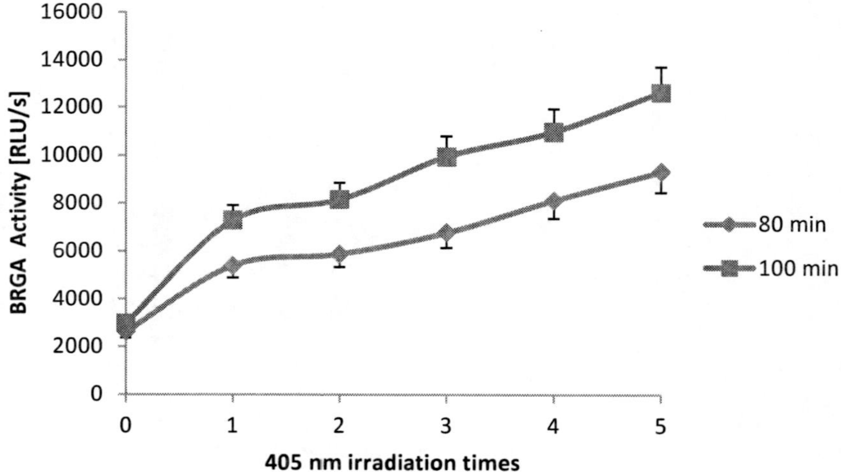

Figure 5. Kinetic of blood ROS generation in 405 nm excited blood after 100 min pre-incubation. The results of figure 4 were reproduced with 80 min or 100 min incubation time. 5-fold 405 nm excitation by the PHOmo microplate photometer resulted in about 4-fold increased blood ROS generation assay (BRGA) activity, determined in relative light units per second (RLU/s).

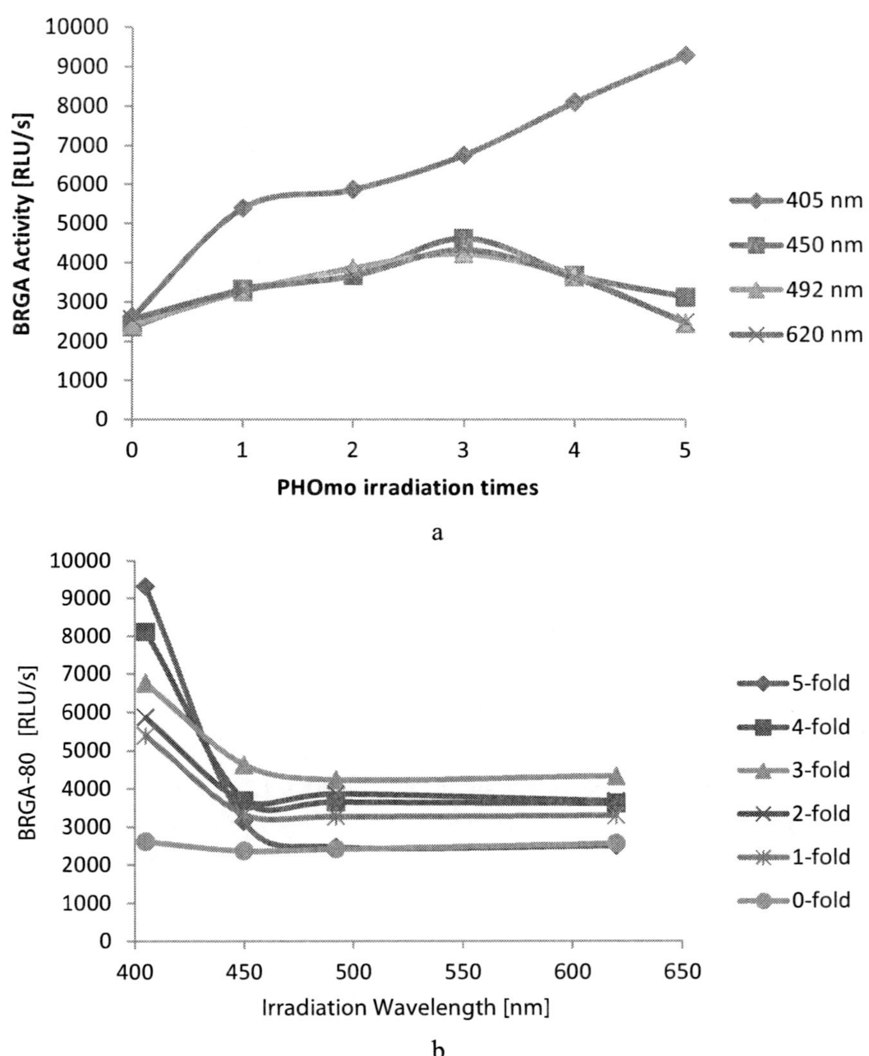

Figure 6. Wavelength dependence of the increase in blood ROS generation. The experiment of figure 4 with 80 min incubation time (BRGA-80) was performed with irraditions at the wavelengths 405 nm (violet), 450 nm (blue), 492 nm (green), or 620 nm (orange). 3-fold excitation with a wavelength $\geq$ 450 nm could double ZyA-stimulated blood ROS generation.

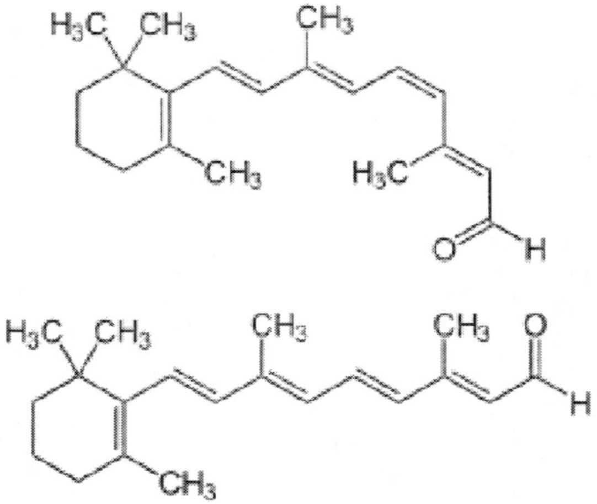

Figure 7. Chemical structure of 11-cis-retinal (above) and all-trans-retinal (below) [11]. A photon hits at the fifth C-atom according to the terminology that the C1 is the highest oxidized C-atom, here the aldehyde of retinal. The energy caught by the second C=C alkene group is sufficient to switch the molecule from the *cis* into the *trans* transformation.

Discussion

The neutrophil´s ROS generation is usually either triggered by compounds that directly bind to the cell´s CD11b/18 receptors in the membrane [11,50-56] or that directly activate protein kinase C (PKC). Zymosan A is a typical activator of CD11b/18, phorbol myristate acetate (PMA) is a typical activator of PKC. For the BRGA the pathophysiological substance ZyA is the trigger of choice, PMA is only a secondarily used trigger. Every compound that only prepares CD11b/18 or PKC for specific activation is defined as a primer. Both, triggers and primers could be considered as stimulators of blood ROS generation, the trigger being a much greater stimulator than the primer. There are few specific triggers but many primers; here evidence is presented that violet photons prime neutrophils for assembly of active NADPH-oxidase.

The neutrophils perceive violet light, in the darkness of the blood stream with reddish erythrocytes it seems plausible that they communicate with each other by violet light quants (the energy of a photon $E=hv$, h being the Planck constant (quantum of action) v (spoken: nü) being the frequency of the photon´s associated electromagnetic wave, the lower the wave length the higher the energy of the photon) [7]. Violet photons "awaken the sleeping" neutrophil and alert (prime) the cell to have its weapons prepared just in case that there are dangerous ROS generation - triggering enemies, such as bacteria, fungi, or thrombi [57-60]. Priming could be a cytoskeletal builder, constructing "highways" for urgent transports of cell proteins. White light of the non-violet spectrum also has a slight priming action, however, too much of it damage the specific violet light receptors [61]. The question arises: what are the eyes of our neutrophils [7]? Known proteins that perceive light signals are the opsins and the flavoproteins. The photon receptor of the opsins is 11-*cis*-retinal that upon reaction of the C_{11} atom at the second C=C with a photon converts into all-*trans*-retinal (Figure 1), the changed 11-*cis*-retinal/ all-*trans*-retinal leaves opsin, specific ion channels activate the cell. The flavoproteins (e.g. cryptochromes) are other candidates for specific receptors of violet/blue light [62-67].

Molecular oxygen (O_2) has two reactive electrons with parallel spin (spin up – spin up) in two anti-selfbonding 2p-orbitals. Excited singlet molecular oxygen (${}^1\Delta O_2^*$) has two highly reactive electrons with anti-parallel spin (spin up - spin down) in a single one of the two anti-selfbonding 2p-orbitals, the other one is empty [11]. ${}^1\Delta O_2^*$ can live for about 1h [68]. ${}^1\Delta O_2^*$ can decay in monomol emission at 1270 nm (invisible), in ${}^1\Delta O_2^*\,{}^1\Delta O_2^*$ dimol emission at 634 nm (red spectrum), or it can react with R_1-C=C-R_2 to excited carbonyls (C=O*) that mostly decay at about 400-425 nm (violet/blue spectrum, the exact wavelength depends on R_1 or R_2 and on the matrix of the ambience (7,11,14,23,24,69)). This could explain the contradictory findings on unsaturated, especially polyunsaturated fatty acids (PUFA), modulators of membranous lipid rafts [70-73]: the cell signal ${}^1\Delta O_2^*$ reacts with PUFA generating violet photons [74-84], the tissue destructor ${}^1\Delta O_2^*$ is inactivated (quenched) by PUFA [85-88].

The biological action of PUFA is also complicated because they are CYP-450 substrates [89], CYP-450 metabolites might act as redox-cyclers, i.e. generator of ${}^1\Delta O_2^*$ that might prime blood neutrophils [14,90-93]. The PUFA might also quench physiological (signal) or pathological (ultraviolet) photons [94].

In conclusion, violet/blue photons of about 400-425 nm prime neutrophils for ROS generation. This is of importance in the physiological and patho-physiological understanding of inflammation. Physiological inflame-mation as e.g. in thrombosis [95] could be stimulated by intra-venous "injection" of violet/blue light, in pathological inflammation violet/blue photon generation or action must be inhibited [96-119].

ACKNOWLEDGMENTS

No specific funding and no conflicts of interest.

REFERENCES

[1] Stief TW. The Laboratory Diagnosis of the Pre-Phase of Pathologic Disseminated Intravascular Coagulation; In: *Handbook of Hematology Research: Hemorheology, Hemophilia and Blood Coagulation*; Tondre R, Lebègue C, eds; Nova Science Publishers; New York; 2011; pp. 255-270.

[2] Stief TW. Thrombin – applied clinical biochemistry of the main factor of coagulation In: *Thrombin: function and pathophysiology*. Stief T, ed.; Nova Science Publishers; New York; 2012; pp. vii-xx. https://www.novapublishers.com/catalog/product_info.php?products_id=33386

[3] Collen D, Lijnen HR. The fibrinolytic system in man. *Crit Rev Oncol Hematol*. 1986; 4: 249-301.

[4] Stief TW. Neutrophil Granulocytes in Hemostasis. In: *Handbook of Hematology Research: Hemorheology, Hemophilia and Blood Coagulation*; Tondre R, Lebègue C, eds; Nova Science Publishers; New York; 2011; pp. 179-200.

[5] Sitrin RG, Pan PM, Harper HA, Todd RF 3rd, Harsh DM, Blackwood RA. Clustering of urokinase receptors (uPAR; CD87) induces proinflammatory signalling in human polymorphonuclear neutrophils. *J Immunol*. 2000; 165: 3341-9.

[6] Syrovets T, Lunov O, Simmet T. Plasmin as a proinflammatory cell activator. *J Leukoc Biol*. 2012; 92: 509-19.

[7] Stief TW. The blood fibrinolysis / deep-sea analogy: a hypothesis on the cell signals singlet oxygen/photons as natural antithrombotics. *Thromb Res*. 2000; 99: 1-20.

[8] Cao D, Mizukami IF, Garni-Wagner BA, Kindzelskii AL, Todd RF 3rd, Boxer LA, Petty HR. Human urokinase-type plasminogen activator primes neutrophils for superoxide anion release. Possible roles of complement receptor type 3 and calcium. *J Immunol*. 1995; 154: 1817-29.

[9] Kindzelskii AL, Laska ZO, Todd RF 3rd, Petty HR. Urokinase-type plasminogen activator receptor reversibly dissociates from complement

receptor type 3 (CD11b/CD18) during neutrophil polarization. *J Immunol*. 1996; 156: 297-309.

[10] Nygren H, Eriksson C, Lausmaa J. Adhesion and activation of platelets and polymorphonuclear granulocyte cells at TiO_2 surfaces. *J Lab Clin Med*. 1997; 129: 35-46.

[11] www.wikipedia.org

[12] Bae YB, Oh H, Rhee SG, Yoo YD. Regulation of reactive oxygen species generation in cell signalling. *Mol. Cells*. 2011; 32: 491-509.

[13] Yasui H, Hayashi S, Sakurai H. Possible involvement of singlet oxygen as multiple oxidants in p450 catalytic reactions. *Drug Metab Pharmacokinet*. 2005; 20: 1-13.

[14] Stief TW. Hemostasis tolerable singlet oxygen - a perspective in AIDS therapy. *Hemost Lab* 2008; 1: 21-40.

[15] Czesnikiewicz-Guzik M, Lorkowska B, Zapala J, Czajka M, Szuta M, Loster B, Guzik TJ, Korbut R. NADPH oxidase and uncoupled nitric oxide synthase are major sources of reactive oxygen species in oral squamous cell carcinoma. Potential implications for immune regulation in high oxidative stress conditions. *J Physiol Pharmacol*. 2008; 59: 139-52.

[16] Sies H, de Groot H. Role of reactive oxygen species in cell toxicity. *Toxicol Lett*. 1992; 64-65: 547-51.

[17] Klotz LO, Kröncke KD, Sies H. Singlet oxygen-induced signaling effects in mammalian cells. *Photochem Photobiol Sci*. 2003; 2: 88-94.

[18] Briviba K, Klotz LO, Sies H. Toxic and signaling effects of photochemically or chemically generated singlet oxygen in biological systems. *Biol Chem*. 1997; 378: 1259-65.

[19] Parke DV, Sapota A. Chemical toxicity and reactive oxygen species. *Int J Occup Med Environ Health*. 1996; 9: 331-40.

[20] Stief T. Micro-thrombi stimulate blood ROS generation. *Hemost Lab*. 2013; 6 (issue 2-3)

[21] Petrone WF, English DK, Wong K, McCord JM. Free radicals and inflammation: superoxide-dependent activation of a neutrophil chemotactic factor in plasma. *Proc Natl Acad Sci USA*. 1980; 77: 1159-63.

[22] Stief TW. Nonradical excited oxygen species induce selective thrombolysis in vivo. *Thromb Res*. 1991; 62: 147-63.

[23] Stief TW. The physiology and pharmacology of singlet oxygen. *Med Hypoth*. 2003; 60: 567-72.

[24] Li Q, Syrovets T, Simmet T, Ding J, Xu J, Chen W, Zhu D, Gao P. Plasmin induces intercellular adhesion molecule 1 expression in human endothelial cells via nuclear factor – kB/mitogen-activated protein kinases-dependent pathways. *Exp Biol Med (Maywood)* 2013; 238: 176-86.

[25] Stief T. Cortisol's suppression of blood ROS generation quantified. *Hemostasis Laboratory* 2013; 6 (issue 2-3).

[26] Del Maestro R, Thaw HH, Björk J, Planker M, Arfors KE. Free radicals as mediators of tissue injury. *Acta Physiol Scand Suppl*. 1980; 492: 43-57.

[27] Sluiter W, de Vree WJ, Pietersma A, Koster JF. Prevention of late lumen loss after coronary angioplasty by photodynamic therapy: role of activated neutrophils. *Mol Cell Biochem*. 1996; 157: 233-8.

[28] Plaetzer K, Krammer B, Berlanda J, Berr F, Kiesslich T. Photophysics and photochemistry of photodynamic therapy: fundamental aspects. *Lasers Med Sci*. 2009; 24: 259-68.

[29] Verhille M, Couleaud P, Vanderesse R, Brault D, Barberi-Heyob M, Froschot C. Modulation of photosensitization processes for an improved targeted photodynamic therapy. *Curr Med Chem*. 2010; 17: 3925-43.

[30] Stief TW. A direct approach in fibrinolysis diagnostic: mimicry of the leukocyte attack by oxidants. *Thromb Res*. 1989; 56: 213-20.

[31] Stief TW. Oxidized fibrin stimulates the activation of pro-urokinase and is the preferential substrate of human plasmin. *Blood Coagul Fibrinolysis* 1993; 4: 117-22.

[32] Stief TW, Kurz J, Doss MO, Fareed J. Singlet oxygen ($^1O_2$) inactivates fibrinogen, factor V, factor VIII, factor X, and platelet aggregation of human blood. *Thromb Res*. 2000; 97: 473-80.

[33] Stief TW, Fareed J. The antithrombotic factor singlet oxygen/light ($^1O_2$/hv). *Clin Appl Thrombosis/Hemostasis* 2000; 6: 22-30.

[34] Stief TW, Kropf J, Kretschmer V, Doss MO, Fareed J. Singlet oxygen ($^1O_2$) inactivates plasmatic free and complexed α_2-macroglobulin. *Thromb Res*. 2000; 98: 541-47.

[35] Stief TW, Jeske WP, Walenga J, Schultz C, Kretschmer V, Fareed J. Singlet oxygen inhibits agonist-induced P-selectin expression and formation of platelet aggregates. *Clin Appl Thrombosis/Hemostasis* 2001; 7: 219-24.

[36] Stief TW, Feek U, Ramaswamy A, Kretschmer V, Renz H, Fareed J. Singlet oxygen ($^1O_2$) disrupts platelet aggregates. *Thromb Res*. 2001; 104/5: 361-9.

[37] Stief TW. NADPH-Oxidase might also act as an antithrombotic. *Circulation* 2003; 107:e8

[38] Stief TW, Fröhlich S, Renz H. Determination of the global fibrinolytic state. *Blood Coagulation and Fibrinolysis* 2007; 18: 479-87.

[39] Stief TW. Singlet oxygen potentiates thrombolysis. *Clin Appl Thrombosis/Hemostasis* 2007; 13: 259-78.

[40] Stief TW. Singlet oxygen enhances intrinsic thrombolysis. The intrinsic oxidative clot lysis assay (INOXCLA). *Clin Appl Thrombosis/ Hemostasis* 2007; 13: 369-83.

[41] Stief TW. Modulation of granulocyte mediated thrombolysis. *Hemostasis Laboratory* 2008; 1: 77-102.

[42] Stief TW. Singlet oxygen and thrombin generation: 0.5 - 1 mM chloramine as anti-viral therapy. *Hemostasis Laboratory* 2010; 3: 311-24.

[43] Stief TW. Singlet oxygen and thrombin generation in RECA. *Hemostasis Laboratory* 2011; 4: 237-54.

[44] Stief T. The routine blood ROS generation assay (BRGA) triggered by typical septic concentrations of zymosan A. *Hemostasis Laboratory* 2013; 6: 89-98.

[45] Stief T. Glucose initially inhibits and later stimulates blood ROS generation. *J Diabetes Mellitus* 2013; 3: 15-21.

[46] Stief TW, Max M. Active endotoxin in sepsis. *Hemostasis Laboratory* 2008; 1: 53-60.

[47] Stief TW. Fungal sepsis can strongly increase plasmatic procalcitonin concentration. *Hemost Lab* 2012; 5: 27-33.

[48] Stief TW. Rapid diagnosis of fungal sepsis by the oxidative *Limulus* test. *Hemost Lab* 2010; 3: 289-96.

[49] Freitas M, Porto G, Lima JL, Fernandes E. Optimization of experimental settings for the analysis of human neutrophils oxidative burst in vitro. *Talanta*. 2009; 78: 1476-83.

[50] Mambula SS, Simons ER, Hastey R, Selsted ME, Levitz SM. Human neutrophil-mediated nonoxidative antifungal activity against *Cryptococcus neoformans*. *Infect Immun.* 2000; 68: 6257-64.

[51] Vetvicka V, Dvorak B, Vetvickova J, Richter J, Krizan J, Sima P, Yvin JC. Orally administered marine (1→3)-ß-D-glucan Phycarine stimulates both humoral and cellular immunity. *Int J Biol Macromolec.* 2007; 40: 291-8.

[52] Soloviev DA, Jawhara S, Fonzi WA. Regulation of innate immune response to Candida albicans infections by $\alpha_M ß_2$-Pra 1p interaction. *Infect Immun*. 2011; 79: 1546-58.

[53] Wright SD, Weitz JI, Huang AJ, Levin SM, Silverstein SC, Loike JD. Complement receptor type three (CD11b/CD18) of human polymorphonuclear leukocytes recognizes fibrinogen. *Proc Natl Acad Sci USA*. 1988; 85: 7734-8.

[54] Loike JD, Silverstein R, Wright SD, Weitz JI, Huang AJ, Silverstein SC. The role of protected extracellular compartments in interactions between leukocytes, and platelets, and fibrin/fibrinogen matrices. *Ann NY Acad Sci* 1992; 667: 163-72.

[55] Loike JD, el Khoury J, Cao L, Richards CP, Rascoff H, Mandeville JT, Maxfield FR, Silverstein SC. Fibrin regulates neutrophil migration in response to interleukin 8, leukotriene B4, tumor necrosis factor, and formyl-methionyl-leucyl-phenylalanine. *J Exp Med* 1995; 181: 1763-72.

[56] Egesten A, Gullberg U, Olsson I, Richter J. Phorbol ester-induced degranulation in adherent human eosinophil granulocytes is dependent on CD11/CD18 leukocyte integrins. *J Leukoc Biol*. 1993; 53: 287-93.

[57] Smith SO. Insights into the activation mechanism of the visual receptor rhodopsin. *Biochem Soc Trans*. 2012; 40: 389-93.

[58] Jacobs GH. The evolution of vertebrate color vision. *Adv Exp Med Biol*. 2012; 739:156-72.

[59] Ou WB, Yi T, Kim JM, Khorana HG. The roles of transmembrane domain helix-III during rhodopsinphotoactivation. *PLoS One*. 2011; 6: e17398.

[60] Ahuja S, Crocker E, Eilers M, Hornak V, Hirshfeld A, Ziliox M, Syrett N, Reeves PJ, Khorana HG, Sheves M, Smith SO. Location of the retinal chromophore in the activated state of rhodopsin*. *J Biol Chem*. 2009; 284: 10190-201.

[61] McKibbin C, Farmer NA, Jeans C, Reeves PJ, Khorana HG, Wallace BA, Edwards PC, Villa C, Booth PJ. Opsin stability and folding: modulation by phospholipid bicelles. *J Mol Biol*. 2007; 374: 1319-32.

[62] Lukacs A, Haigney A, Brust R, Zhao RK, Stelling AL, Clark IP, Towrie M, Greetham GM, Meech SR, Tonge PJ. Photoexcitation of the blue light using FAD photoreceptor AppA results in ultrafast changes to the protein matrix. *J Am Chem Soc*. 2011; 133: 16893-900.

[63] Wolf MM, Zimmermann H, Diller R, Domratcheva T. Vibrational mode analysis of isotope-labeled electronically excited riboflavin. *J Phys Chem B*. 2011; 115: 7621-8.

[64] Clark IP, George MW, Greetham GM, Harvey EC, Long C, Manton JC, Pryce MT. Photochemistry of (η(6)-arene)Cr(CO)3 (arene = methylbenzoate, naphthalene, or phenanthrene) in n-heptane solution: population of two excited states following 400 nm excitation as detected by picosecond time-resolved infrared spectroscopy. *J Phys Chem A*. 2011; 115: 2985-93.

[65] Haigney A, Lukacs A, Zhao RK, Stelling AL, Brust R, Kim RR, Kondo M, Clark I, Towrie M, Greetham GM, Illarionov B, Bacher A, Römisch-Margl W, Fischer M, Meech SR, Tonge PJ. Ultrafast infrared spectroscopy of an isotope-labeled photoactivatable flavoprotein. *Biochemistry*. 2011; 50: 1321-8.

[66] Krauss U, Minh BQ, Losi A, Gärtner W, Eggert T, von Haeseler A, Jaeger KE. Distribution and phylogeny of light-oxygen-voltage-blue-light-signaling proteins in the three kingdoms of life. *J Bacteriol.* 2009 191: 7234-42.

[67] Losi A, Gärtner W. Old chromophores, new photoactivation paradigms, trendy applications: flavins in blue light-sensing photoreceptors. *Photochem Photobiol.* 2011; 87: 491-510.

[68] Wilkinson F, Helman WP, Ross AB. Rate constants for the decay and reactions of the lowest electronically excited singlet state of molecular oxygen in solution. An expanded and revised compilation. *Journal of Physical and Chemical Reference Data* 1995; 24: 663-77.

[69] Stief TW. Regulation of hemostasis by singlet oxygen ($^1\Delta O_2$). *Curr Vasc Pharmacol* 2004; 2: 357-62.

[70] Shaikh SR. Biophysical and biochemical mechanisms by which dietary N-3 polyunsaturated fatty acids from fish oil disrupt membrane lipid rafts. *J Nutr Biochem*. 2012; 23: 101-5.

[71] Shaikh SR, Jolly CA, Chapkin RS. n-3 Polyunsaturated fatty acids exert immunomodulatory effects on lymphocytes by targeting plasma membrane molecular organization. *Mol Aspects Med*. 2012; 33: 46-54.

[72] McMurray DN, Bonilla DL, Chapkin RS. n-3 Fatty acids uniquely affect anti-microbial resistance and immune cell plasma membrane organization. *Chem Phys Lipids*. 2011; 164: 626-35.

[73] Guichardant M, Chen P, Liu M, Calzada C, Colas R, Véricel E, Lagarde M. Functional lipidomics of oxidized products from polyunsaturated fatty acids. *Chem Phys Lipids*. 2011; 164: 544-8.

[74] Moreno C, Macias A, Prieto A, De La Cruz A, Valenzuela C. Polyunsaturated fatty acids modify the gating of kv channels. *Front Pharmacol*. 2012; 3: 163.

[75] Mrakic-Sposta S, Gussoni M, Montorsi M, Porcelli S, Vezzoli A. Assessment of a standardized ROS production profile in humans by electron paramagnetic resonance. *Oxid Med Cell Longev*. 2012: 973927.

[76] Prasad A, Pospíšil P. Linoleic acid-induced ultra-weak photon emission from Chlamydomonas reinhardtii as a tool for monitoring of lipid peroxidation in the cell membranes. *PLoS One*. 2011; 6: e22345.

[77] Enriquez MM, Fuciman M, LaFountain AM, Wagner NL, Birge RR, Frank HA. The intramolecular charge transfer state in carbonyl-containing polyenes and carotenoids. *J Phys Chem B*. 2010; 114: 12416-26.

[78] Velosa AC, Baader WJ, Stevani CV, Mano CM, Bechara EJ. 1,3-diene probes for detection of triplet carbonyls in biological systems. *Chem Res Toxicol*. 2007; 20: 1162-9.

[79] Feenstra JS, Park ST, Zewail AH. Excited state molecular structures and reactions directly determined by ultrafast electron diffraction. *J Chem Phys*. 2005; 123: 221104.

[80] Cilento G, Adam W. From free radicals to electronically excited species. *Free Radic Biol Med*. 1995; 19: 103-14.

[81] Nascimento AL, Cilento G. Chemiexcitation in the arachidonic acid cascade. *Photochem Photobiol*. 1991; 53: 379-84.

[82] Barnard ML, Gurdian S, Turrens JF. Activated polymorphonuclear leukocytes increase low-level chemiluminescence of isolated perfused rat lungs. *J Appl Physiol*. 1993; 75: 933-9.

[83] Chábera P, Fuciman M, Hríbek P, Polívka T. Effect of carotenoid structure on excited-state dynamics of carbonylcarotenoids. *Phys Chem Chem Phys*. 2009; 11: 8795-803.

[84] Shanati A, Rivlin Y, Shnizer S, Rosenschein U, Goldhammer E. Serum oxidizability potential of ischemic heart disease patients is associated with exercise test results and disease severity. *World J Cardiol*. 2009; 1: 46-50.

[85] Smith DA. Review: Omega-3 polyunsaturated fatty acid supplements do not reduce major cardiovascular events in adults. *Ann Intern Med*. 2012; 157: JC6-5.

[86] Kazemian P, Kazemi-Bajestani SM, Alherbish A, Steed J, Oudit GY. The use of ω-3 poly-unsaturated fatty acids in heart failure: a preferential role in patients with diabetes. *Cardiovasc Drugs Ther*. 2012; 26: 311-20.

[87] Abeywardena MY, Patten GS. Role of ω3 long-chain polyunsaturated fatty acids in reducing cardio-metabolic risk factors. *Endocr Metab Immune Disord Drug Targets*. 2011; 11: 232-46.

[88] Hull MA. Omega-3 polyunsaturated fatty acids. *Best Pract Res Clin Gastroenterol*. 2011; 25: 547-54.

[89] Konkel A, Schunck WH. Role of cytochrome P450 enzymes in the bioactivation of polyunsaturated fatty acids. *Biochim Biophys Acta*. 2011; 1814: 210-22.

[90] Kitada M, Horie T, Awazu S. Chemiluminescence associated with doxorubicin-induced lipid peroxidation in rat heart mitochondria. *Biochem Pharmacol*. 1994; 48: 93-9.

[91] Weiss SJ, Lampert MB, Test ST. Long-lived oxidants generated by human neutrophils: characterization and bioactivity. *Science*. 1983; 222: 625-8.

[92] Stief TW. Singlet oxygen – oxidizable lipids in the HIV membrane, new targets for AIDS therapy ? *Med Hypoth* 2003; 60: 575-7.

[93] Stief TW, Kurz J. The natural anti-inflammatory agent PAI-2 suppresses the oxidative state of human blood. *Hemostasis Laboratory* 2012; 5: 145-59.

[94] Pilkington SM, Watson RE, Nicolaou A, Rhodes LE. Omega-3 polyunsaturated fatty acids: photoprotective macronutrients. *Exp Dermatol*. 2011; 20: 537-43.

[95] Stief TW. Hemostasis and infection/inflammation. *Hemostasis Laboratory* 2008; 1: 1-2.

[96] Bagga B, Pahuja S, Murthy S, Sangwan VS. Endothelial failure after collagen cross-linking with riboflavin and UV-A: case report with literature review. *Cornea*. 2012; 31: 1197-200.

[97] Duchatelle V, Kritikou EA, Tardif JC. Clinical value of drugs targeting inflammation for the management of coronary artery disease. *Can J Cardiol*. 2012 ; 28: 678-86.

[98] Zaba LC, Fiorentino DF. Skin disease in dermatomyositis. *Curr Opin Rheumatol*. 2012 ; 24: 597-601.

[99] Caorsi R, Federici S, Gattorno M. Biologic drugs in autoinflammatory syndromes. *Autoimmun Rev*. 2012; 12: 81-6.

[100] Huang YY, Gupta A, Vecchio D, Arce VJ, Huang SF, Xuan W, Hamblin MR. Transcranial low level laser (light) therapy for traumatic brain injury. *J Biophotonics*. 2012; 5: 827-37.

[101] Komatsu N, Takayanagi H. Autoimmune arthritis: the interface between the immune system and joints. *Adv Immunol*. 2012; 115:45-71.

[102] Kijlstra A, Tian Y, Kelly ER, Berendschot TT. Lutein: more than just a filter for blue light. *Prog Retin Eye Res*. 2012; 31: 303-15.

[103] Ozawa Y, Sasaki M, Takahashi N, Kamoshita M, Miyake S, Tsubota K. Neuroprotective effects of lutein in the retina. *Curr Pharm Des*. 2012; 18: 51-6.

[104] Nag TC, Wadhwa S. Ultrastructure of the human retina in aging and various pathological states. *Micron*. 2012; 43: 759-81.

[105] Hjorth E, Freund-Levi Y. Immunomodulation of microglia by docosahexaenoic acid and eicosapentaenoic acid. *Curr Opin Clin Nutr Metab Care*. 2012; 15: 134-43.

[106] Nowsheen S, Aziz K, Kryston TB, Ferguson NF, Georgakilas A. The interplay between inflammation and oxidative stress in carcinogenesis. *Curr Mol Med*. 2012; 12: 672-80.

[107] Stief TW. Xenobiotic - induced pancreas carcinoma following mesenteric vein thrombosis : a hypothesis. *Hemostasis Laboratory* 2010; 3: 253-8. https://www.novapublishers.com/catalog/product_info.php? products_id=27631

[108] Ullrich SE, Byrne SN. The immunologic revolution: photoimmunology. *J Invest Dermatol*. 2012; 132 (3 Pt 2): 896-905.

[109] Chung H, Dai T, Sharma SK, Huang YY, Carroll JD, Hamblin MR. The nuts and bolts of low-level laser (light) therapy. *Ann Biomed Eng*. 2012; 40: 516-33.

[110] Hinterdorfer P, Garcia-Parajo MF, Dufrêne YF. Single-molecule imaging of cell surfaces using near-field nanoscopy. *Acc Chem Res*. 2012; 45: 327-36.

[111] Vezzani A, Aronica E, Mazarati A, Pittman QJ. Epilepsy and brain inflammation. *Exp Neurol*. 2013; 244: 11-21.

[112] Shores DR, Binion DG, Freeman BA, Baker PR. New insights into the role of fatty acids in the pathogenesis and resolution of inflammatory bowel disease. *Inflamm Bowel Dis*. 2011; 17: 2192-204.

[113] Serini S, Fasano E, Piccioni E, Cittadini AR, Calviello G. Dietary n-3 polyunsaturated fatty acids and the paradox of their health benefits and potential harmful effects. *Chem Res Toxicol*. 2011; 24: 2093-105.

[114] Anders HJ, Ryu M. Renal microenvironments and macrophage phenotypes determine progression or resolution of renal inflammation and fibrosis. *Kidney Int*. 2011; 80: 915-25.

[115] Zindler E, Zipp F. Neuronal injury in chronic CNS inflammation. *Best Pract Res Clin Anaesthesiol*. 2010; 24: 551-62.

[116] Nicolaou A, Pilkington SM, Rhodes LE. Ultraviolet-radiation induced skin inflammation: dissecting the role of bioactive lipids. *Chem Phys Lipids*. 2011; 164: 535-43.
[117] Brightling CE, Gupta S, Gonem S, Siddiqui S. Lung damage and airway remodelling in severe asthma. *Clin Exp Allergy*. 2012 ; 42: 638-49.
[118] Homer RJ, Elias JA, Lee CG, Herzog E. Modern concepts on the role of inflammation in pulmonary fibrosis. *Arch Pathol Lab Med.* 2011; 135: 780-8.
[119] Saffar AS, Ashdown H, Gounni AS. The molecular mechanisms of glucocorticoids-mediated neutrophil survival. *Curr Drug Targets*. 2011; 12: 556-62.

Chapter 2

BLOOD NEUTROPHILS SEE UV LIGHT: 340 NM ULTRAVIOLET A STIMULATES BLOOD ROS GENERATION NEARLY HALF AS STRONG AS 405 NM VIOLET PHOTONS

ABSTRACT

Background: Neutrophils communicate with each other in the dark blood vessel. Light is the only signal in flowing blood whose origin can be recognized immediately. Via NADPH-oxidase assembly and secreted myeloperoxidase the neutrophils generate the excited (*) oxygen species singlet molecular oxygen ($^1\Delta O_2^*$), that reacts with C=C to excited carbonyls (R-C=O*) that emit light quants of a wavelength around 400 nm. Of great importance would be the existence of specific photon receptors that transmit this light information to the cell.

Material and Methods: 10 µl 0.5h old citrated normal blood were added in 4-fold to transparent polystyrene U-microwells (Brand®781600) prefilled with 150 µl Hanks′ Balanced Salt Solution. The blood samples were excited 0-8-fold by light of the wave lengths 405 nm (violet) or 340 nm (ultraviolet) using a microtiter plate photometer (Tecan Sunrise). 100 µl reaction mixture (6.3 µl blood) of the transparent wells were immediately transferred into black polystyrene F-microwells (Brand®781608). 0, 0.8, or 1.5 µg/ml (final) zymosan A (ZyA) as specific ROS generation trigger was added. After addition of 0.4 mM (final) luminol the luminescence (RLU/s) was measured by a microtiter plate luminometer (LUmo) after 0-183 min (37°C). After 100 min (37°C) half of the wells without ROS generation trigger were supplemented with 3 µg/ml ZyA (final) and the luminescence was determined.

Results: 405 nm excitation gave the highest ROS increase. 4fold excitation increased 0.8 μg/ml ZyA-stimulated blood ROS generation up to 10-fold. If excited blood was preincubated for 100 min and then triggered by 3 μg/ml ZyA, the 405 nm excitation again increased the ROS generations. The maximal increase factor was about 4 for 4-fold 405 nm excitation and about 5 for 8-fold 405 nm excitation. 340 nm excitation induced also a strong increase in blood ROS generation; the 340 nm excitation was nearly half as efficient as 405 nm excitation. Without the ZyA-trigger the normal blood ROS generation was always 0 RLU/s, independent of wavelength, excitation time, pre-incubation time, or incubation time.

Discussion: Human neutrophils can sense ultraviolet to dark blue photons of about 340-425 nm, presumably by opsin. These photonic cell signals prime them for ROS generation. This is of great importance in the understanding of physiological or pathological inflammation. Physiological inflammation as e.g. in cellular fibrinolysis could be stimulated by intra-venous "injection" of 400 nm light (3.1 eV per quant), in pathological inflammation ultraviolet/blue photon generation inhibitors /quenchers are indicated.

Keywords: Neutrophils, light quants, photons, 340 nm, ultraviolet, UVA, 405 nm, violet, blue, ROS, cell signal, photon receptor

INTRODUCTION

Neutrophils recruit other neutrophils for physiologic (or pathologic) inflammation amplification [1,2]. In the darkness of the blood vessels light is the only signal whose origin can be recognized immediately [3]. Via activation of specific cell receptors, e.g. the adhesion receptor CD11b/18, the neutrophil assembles NADPH-oxidase to his cell membrane and secrets myeloperoxidase into the micro-ambience [4]. So, the neutrophils generate the excited (*) oxygen species singlet molecular oxygen ($^1\Delta O_2^*$), that reacts with C=C to excited carbonyls (R-C=O*) that emit light quants of a wavelength around 400 nm depending on the chemical nature of R and on the matrix [5-8]. Of great physiologic importance would be the existence of specific photon receptors that transmit this light information inside the cell. The blood neutrophil could then be compared to a sea fish or an amphibian with retinal rods or cones sensible for ultraviolet light [9-32].

Material and Methods

10 µl 0.5h old normal blood anticoagulated with 11 mM Na_3 - citrate, pH 7.4, in polypropylene tube (time to expiry: about 1 year left; Sarstedt, Nümbrecht, Germany) were added in 4-fold using an Eppendorf-multipette® after written informed consent to transparent high quality polystyrene U-microwells (Brand, Wertheim, Germany; article nr. 781600) prefilled with 150 µl Hanks′ Balanced Salt Solution (HBSS; modified without phenol red; SAFC Biosciences-Sigma, Deisenhofen, Germany; article nr. 55037C-1000ML).

The blood samples were excited 0-8-fold by a microtiter plate photometer (Tecan Sunrise, Crailsheim, Germany), using light of the wave lengths 405 nm (violet) or 340 nm (ultraviolet). 100 µl reaction mixture (6.3 µl blood) of the transparent wells were immediately transferred into black high quality polystyrene F-microwells (Brand; article nr. 781608). 5 µl or 10 µl 18 µg/ml zymosan A (ZyA, 0.8 µg/ml or 1.5 µg/ml final; Sigma, Deisenhofen, Germany; article nr. Z-4250-1G, lot nr. 27H0495) in 0.9% NaCl or no specific ROS generation trigger was added. 10 µl 5 mM luminol sodium salt (0.4 mM final; Sigma) in 0.9% NaCl were added and the luminescence in relative light units per second (RLU/s) was measured by a microtiter plate luminometer (LUmo; anthos, Krefeld, Germany) after 0-183 min (37°C).

After 100 min (37°C) pre-incubation half of the wells without ROS generation trigger were supplemented with 10 µl 36 µg/ml ZyA (3 µg/ml final) in 0.9% NaCl and the luminescence was determined after 0-116 min incubation.

The mean values of the 4-fold determinations were calculated. The intra-assay coefficients of variation were less than 10% at > 500 RLU/s and less than 15% at lower luminol enhanced blue light emissions.

HBSS consisted of 185.4 mg/l $CaCl_2 \cdot 2\ H_2O$, 200 mg/l $MgSO_4 \cdot 7\ H_2O$, 400 mg/l KCl, 60 mg/l KH_2PO_4, 350 mg/l $NaHCO_3$, 8000 mg/l NaCl, 90 mg/l Na_2HPO_4, 1000 mg/l glucose, pH 7.0-7.4. Expressed in molarity, the concentrations of the HBSS components are: 1.3 mM Ca^{2+}, 0.8 mM Mg^{2+}, 5.8 mM K^+, 143 mM Na^+, 144 mM Cl^-, 1.6 mM SO_4^{2-}, 0.4 mM $H_2PO_4^-$, 0.6 mM HPO_4^{2-}, 4.2 mM HCO_3^-, 5.6 mM glucose.

The BRGA [8,33]

1. imitates the physiologic ROS generation answer to blood pathogens (bacteria, fungi, parasites, thrombi)
2. imitates the pathophysiologic blood ROS generation in autoimmunity
3. detects globally the most important ROS of blood, i.e. H_2O_2 + $\cdot OH$ + $^1O_2$
4. uses the stable pathophysiologic trigger hydrophilic ZyA
5. is sensitive enough to detect ROS upon blood stimulation by only about 1 µg/ml ZyA (the approximate concentration of initial fungal sepsis [34-36])
6. does not interfere with cell function (using untoxic luminol concentrations)
7. is cheap: without taxes one LUmo (microtiter plate luminometer) costs less than 4500 €
8. uses HBSS, a close-to-physiologic buffered salt solution
9. is easy to standardize
10. works with whole blood
11. is quantitative and can therefore compare the results with the normal range (MV±1SD)
12. is suitable for routine measurements of hundreds of samples within minutes.

RESULTS

The 0.8 µg/ml ZyA-triggered blood ROS generation kinetic in very fresh normal citrated blood demonstrates that 405 nm excitation resulted in higher blood ROS maxima, directly proportional to the number of excitation times: 47 RLU/s without excitation increased up to 94-103 RLU/s upon 4-8-fold excitation (Figure 1a). After the maximum at 56 min all blood ROS generations declined to less than 20 RLU/s at 105 min, followed by small second ROS generation maxima (possibly due to cell stimulation by micro-thrombi [37]) of about 30 RLU/s at 127 min, again followed by the major ROS decline up to 145 min, and then followed by a third increase of blood ROS generation, depending on the times of 405 nm excitation:

11 RLU/s for 0-fold up to 109 RLU/s for 8-fold. This final blood ROS generation increase reflects cellular fibrinolysis [37].

1.5 μg/ml ZyA-stimulated normal blood had similar characteristics in the initial phase of blood ROS generation, with higher respective blood ROS maxima. However, here the final blood ROS generation increase reflecting cellular fibrinolysis was negatively influenced by 405 nm excitation, presumably because the higher concentration of ZyA competes with fibrin for the specific adhesion receptor CD11b/18 (Figure 1b).

Figure 2 demonstrates the blood ROS generation at the time of the maximum, figure 3 in the first NADPH-oxidase down-regulation phase, and figure 4 in the full phase of cellular fibrinolysis. Always the 0.8 μg/ml ZyA-stimulated blood had the highest ROS increase following 405 nm excitation. 4-fold excitation could increase blood ROS generation up to 10-fold when compared with the unexcited controls, especially pronounced for the low conc. ZyA – trigger. If excited blood was preincubated for 100 min and then triggered by 3 μg/ml ZyA, the 405 nm excitation again increased the considerably higher blood ROS generations (due to preanalytic CD11b/18 priming) (Figure 5). The maximal increase factor was about 4 for 4-fold 405 nm excitation and about 5 for 8-fold 405 nm excitation (Figure 6).

340 nm excitation induced also a strong increase in blood ROS generation (Figures 7-9). When compared with 405 nm excitation (Figures 2-4), the 340 nm excitation was nearly half as efficient as 405 nm excitation to increase blood ROS generation. Again, 100 min preincubation already resulted in very high basal blood ROS generations (Figure 10). The blood ROS generation increased maximally about 2-fold by 4-fold 340 nm excitation (Figure 11).

Without the ZyA-trigger the normal blood ROS generation was always 0 RLU/s, independent of wavelength, excitation time, pre-incubation time, or incubation time, i.e. light quants do not trigger, they prime neutrophils.

If the blood ROS generation induced by 4-fold 405 nm excitation was set as 100%, then the respective blood ROS generations induced by 4-fold excitation at 340 nm, 450 nm, 490 nm, or 620 nm [8] were 48 %, 4 %, 4 %, or 3 %, respectively.

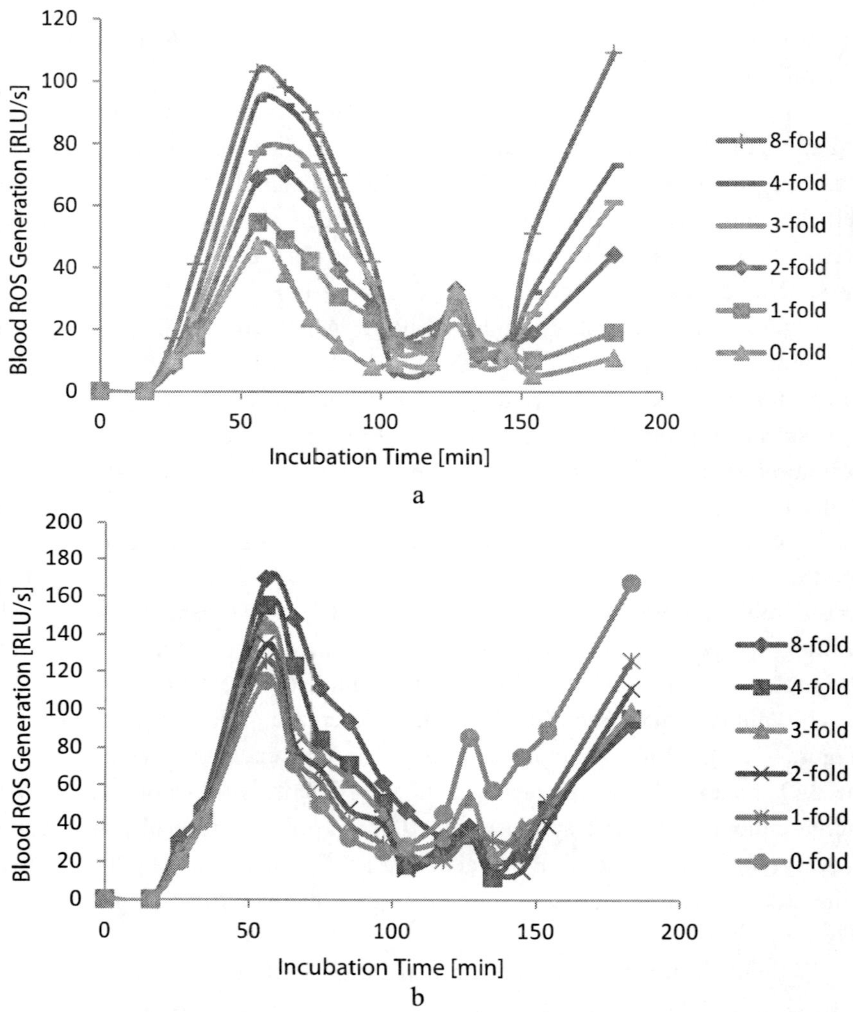

Figure 1. Kinetic of blood ROS generation in 405 nm excitated blood. 10 µl 0.5h old citrated normal blood were added in 4-fold to transparent polystyrene U-microwells (Brand®781600) prefilled with 150 µl Hanks′ Balanced Salt Solution (HBSS). The microtiter plates were excitated at 405 nm 0-fold, 1-fold, 2-fold, 3-fold, 4-fold, or 8-fold by a microtiter plate photometer (Tecan Sunrise). 100 µl reaction mixture of the transparent wells were immediately transferred into black polystyrene F-microwells (Brand®781608). 0.8 µg/ml (Figure 1a) or 1.5 µg/ml zymosan A (Figure 1b; ZyA) and 0.4 mM luminol (final) were added and the luminescence (RLU/s) was measured by a photons multiplying microtiter plate luminometer (LUmo) after 0-183 min (37°C); intra-assay CV values < 15%.

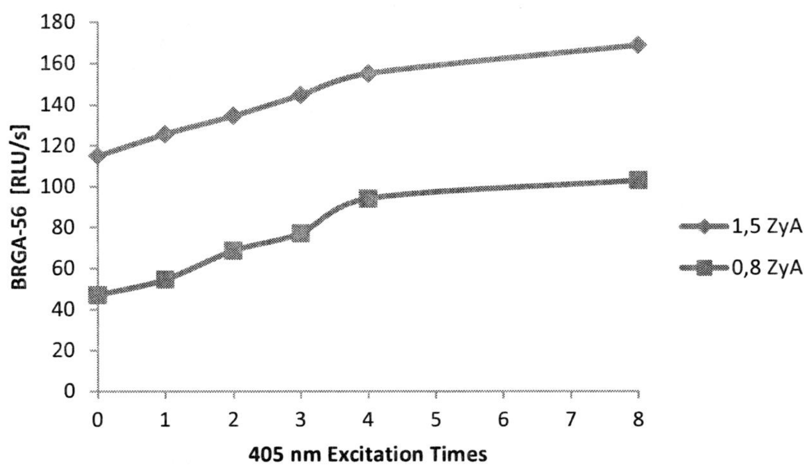

Figure 2. ROS increase by 405 nm excitation. The BRGA activity at 56 min reaction time of figure 1 was plotted against the 405 nm excitation times with the ROS generation in freshest blood being triggered by 0.8 µg/ml ZyA or 1.5 µg/ml ZyA.

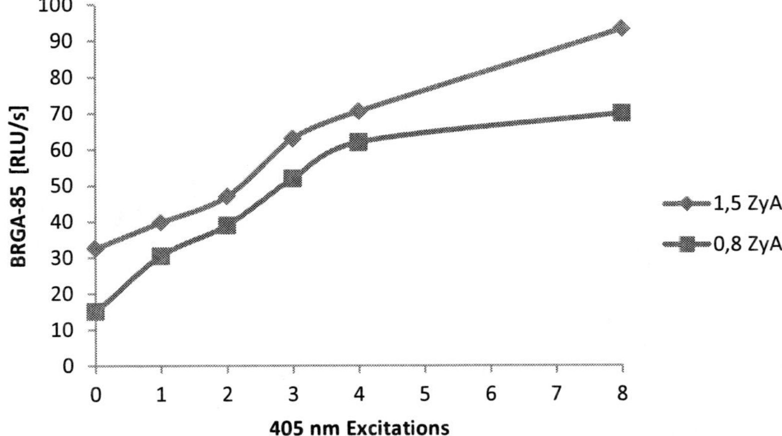

Figure 3. ROS increase by 405 nm excitation. The BRGA activity at 85 min reaction time of figure 1 was plotted against the 405 nm excitation times with the ROS generation in freshest blood being triggered by 0.8 µg/ml ZyA or 1.5 µg/ml ZyA.

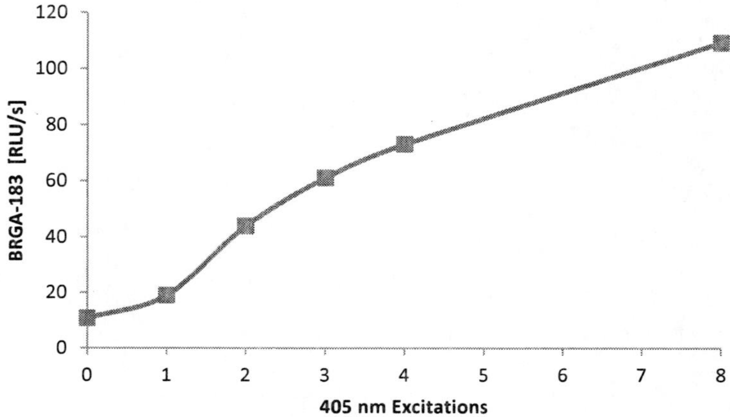

Figure 4. ROS increase by 405 nm excitation. The BRGA activity at 183 min reaction time of figure 1 was plotted against the 405 nm excitation times with the ROS generation in freshest blood being triggered by 0.8 μg/ml ZyA.

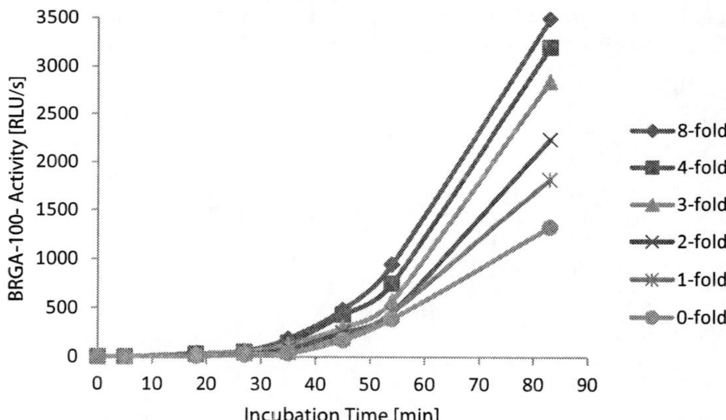

Figure 5. BRGA kinetic (after 100 min preincubation) of blood ROS generation in 405 nm excitated blood. 10 μl 0.5h old citrated normal blood were added in 4-fold to transparent polystyrene U-microwells (Brand®781600) prefilled with 150 μl Hanks′ Balanced Salt Solution (HBSS). The microtiter plates were excitated at 405 nm 0-fold, 1-fold, 2-fold, 3-fold, 4-fold, or 8-fold by a microtiter plate photometer (Tecan Sunrise). 100 μl reaction mixture of the transparent wells were transferred into black polystyrene F-microwells (Brand®781608) and 0.4 mM luminol was added. After 100 min preincuation (37°C) 3 μg/ml ZyA were added and the luminescence (RLU/s) was measured by a photons multiplying microtiter plate luminometer (LUmo) after 0-83 min (37°C); intra-assay CV values < 10%.

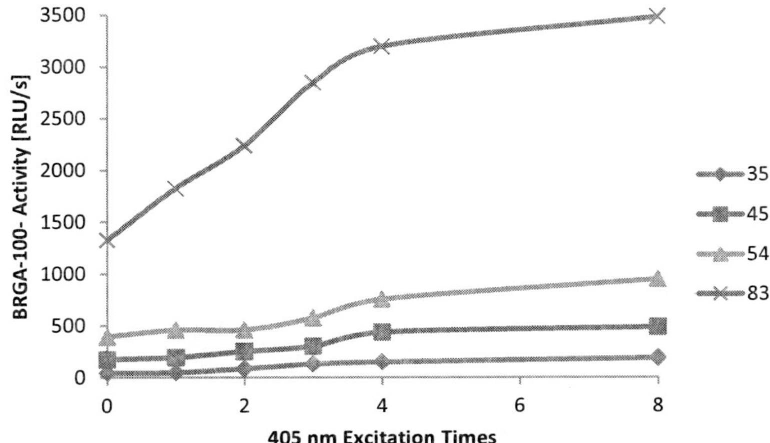

Figure 6. ROS increase by 405 nm excitation. The BRGA activity at 35, 45, 54, or 83 min reaction time of figure 5 was plotted against the 405 nm excitation times with the ROS generation in 100 min preincubated (BRGA-100-) blood being triggered by 3 µg/ml ZyA. The maximal increase factor was about 4 for 4-fold 405 nm excitation and about 5 for 8-fold 405 nm excitation (measured at BRGA-100-35).

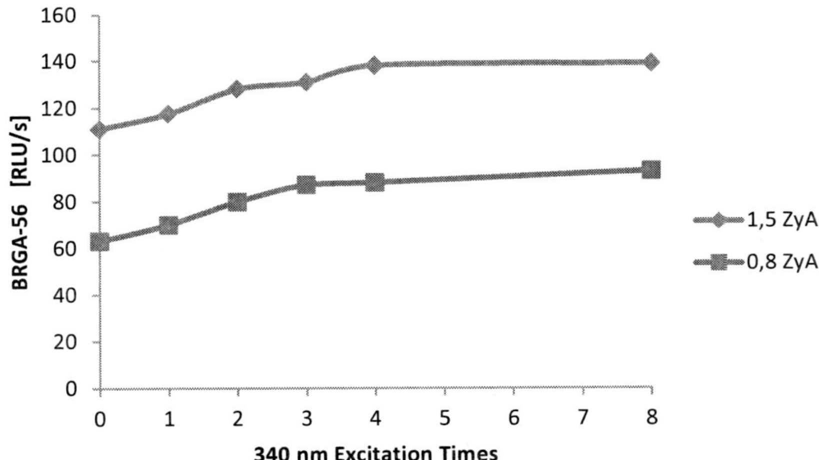

Figure 7. ROS increase by 340 nm excitation. The experiment of figure 1 was performed at 340 nm instead of 405 nm excitation. The BRGA activity at 56 min reaction time of was plotted against the 340 nm excitation times with the ROS generation in freshest blood being triggered by 0.8 µg/ml ZyA or 1.5 µg/ml ZyA.

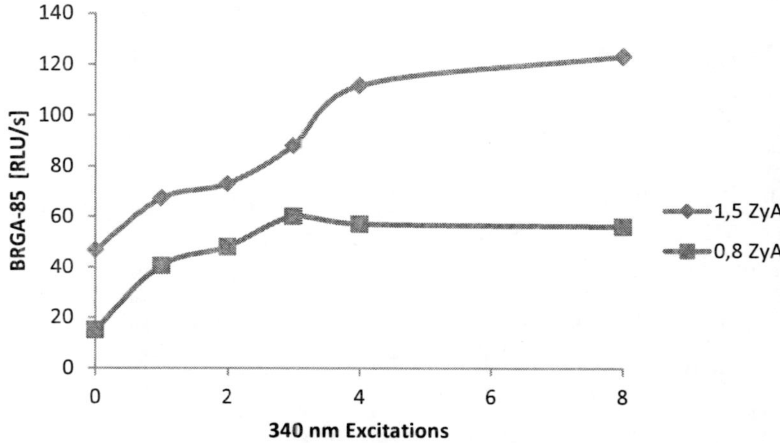

Figure 8. ROS increase by 405 nm excitation. The BRGA activity at 85 min reaction time similar to figure 7 was plotted against the 340 nm excitation times with the ROS generation in freshest blood being triggered by 0.8 μg/ml ZyA or 1.5 μg/ml ZyA.

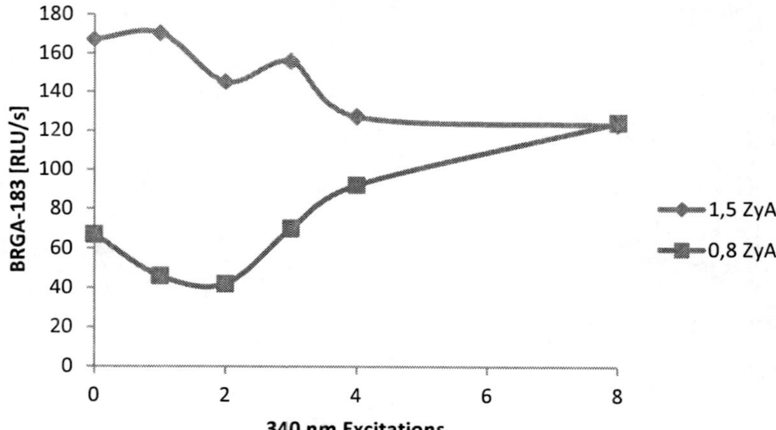

Figure 9. ROS increase by 405 nm excitation. The BRGA activity at 183 min reaction time similar to figure 7 was plotted against the 340 nm excitation times with the ROS generation in freshest blood being triggered by 0.8 μg/ml ZyA or 1.5 μg/ml ZyA.

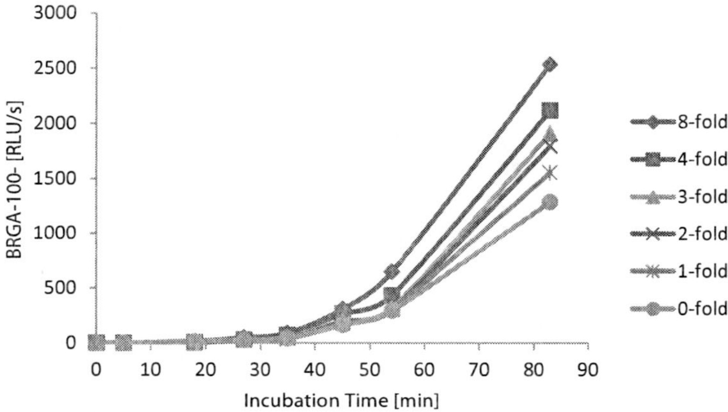

Figure 10. BRGA kinetic (after 100 min preincubation) of blood ROS generation in 340 nm excitated blood. 10 µl 0.5h old citrated normal blood were added in 4-fold to transparent polystyrene U-microwells (Brand®781600) prefilled with 150 µl Hanks′ Balanced Salt Solution (HBSS). The microtiter plates were excitated at 340 nm 0-fold, 1-fold, 2-fold, 3-fold, 4-fold, or 8-fold by a microtiter plate photometer (Tecan Sunrise). 100 µl reaction mixture of the transparent wells were transferred into black polystyrene F-microwells (Brand®781608) and 0.4 mM luminol was added. After 100 min preincuation (37°C) 3 µg/ml ZyA were added and the luminescence (RLU/s) was measured by a photons multiplying microtiter plate luminometer (LUmo) after 0-83 min (37°C); intra-assay CV values < 10%.

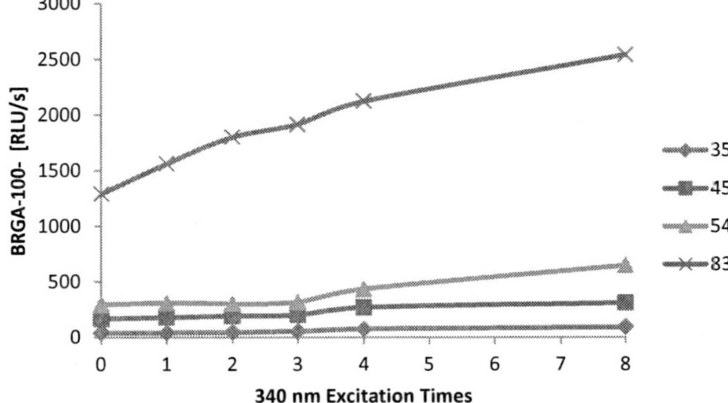

Figure 11. ROS increase by 340 nm excitation. The BRGA activity at 35, 45, 54, or 83 min reaction time of blood samples of figure 10 excited by 340 nm was plotted against the 340 nm excitation times with the ROS generation in 100 min preincubated (BRGA-100-) blood being triggered by 3 µg/ml ZyA. Blood ROS generation increased maximally about 2-fold by 4-fold 340 nm excitation (measured at BRGA-100-35).

DISCUSSION

The present findings have important implications in photons-medicine. The possible uses of photons-medicine are unlimited [38-40]. So, in photodynamic therapy (PDT) a photosensitiser is usually combined with a laser beam. The photosensitiser absorbs the wavelength of the light and produces singlet oxygen, that selectively destroys pathological cells. Either because there is less sensitiser in the adjacent normal tissue or because normal tissue has better repair mechanisms, preferably the pathological tissue is destroyed. Unlike ionising irradiation, repeated injections and treatments can be made indefinitely [38]. The tissue the light has to travel through to get to the photosensitizer weakens the light. The Doherty - photosensitiser porfimer sodium has a peak absorption in the area of 405 nm (violet) and a much lower absorption peak at 630 nm (red). Since 630 nm photons penetrate tissue deeper than 405 nm photons, usually a red laser beam is preferred to excitate the photosensitizer, although 500 nm light stimulates cells better than 600 nm light [41]. However, too strong excitations that generate pathological concentrations of singlet oxygen could result in vessel thrombosis [38].

People born in april tend to have more multiple sclerosis than people born in november [42]. Ultraviolet light seems to induce lupus erythematodes [43]. Light in ultraviolet/violet/blue spectrum seems to be more active than light of higher wavelengths [44-53].

Ultraviolet irradiation is electromagnetic irradiation with a wavelength shorter than that of visible light (380–700 nm), but longer than x-rays (<100 nm) [54]. Medically relevant UV irradiation [54,55] is divided into the spectral areas

UV-A (315–380 nm; 380 nm = 789 THz; 3.94-3.26 eV energy quant per photon).

UV-B (280–315 nm; 4.43-3.94 eV); the peak production of vitamin D occurs at 295-297 nm.

UV-C (200–280 nm; 6.2-4.43 eV); tryptophane (but also tyrosine and phenylalanine) absorb energy at 280 nm; the nucleic acids, especially thymine, absorbs at 265 nm and dimerizes. Peptide bonds are attacked at 220 nm [55]. About 10 eV is needed to tear off an electron out of an atom and to ionize it.

In conclusion, human neutrophils can sense ultraviolet to dark blue photons of about 340-425 nm, possibly by opsin (Figure 13). These photonic cell signals prime them for ROS generation. This is of great importance in the understanding of physiological or pathological inflammation. Physiological inflammation as e.g. in cellular fibrinolysis [56-70] could be stimulated by intra-venous "injection" of 400 nm light (3.1 eV per quant), in pathological inflammation ultraviolet/blue photon generation inhibitors/quenchers are indicated [71].

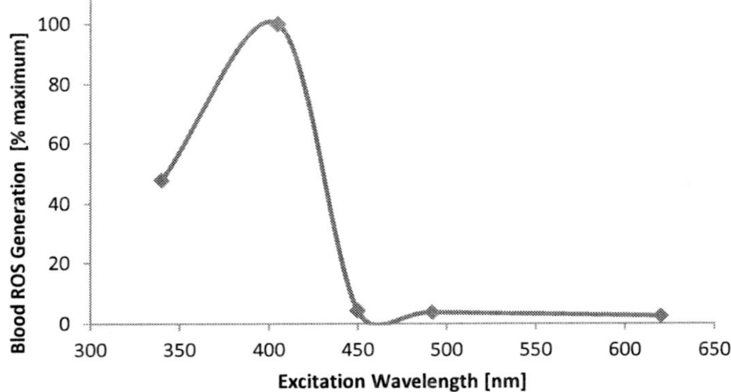

Figure 12. Cell receptor response to violet or ultraviolet photons. The maximal cellular NADPH-oxidase assembly by 4-fold 405 nm excitation was set as 100%. From the 340 nm data of the present work and the 450 nm, 490 nm, 620 nm results of [8] the relative responses of the neutrophil's specific receptor to all these photons were calculated.

Figure 13. Chemical structure of 11-cis-retinal (above) and all-trans-retinal (below) [55]. The C-atom at position 5 of retinal (with the position 1 C-atom being the most oxidized one) captures an ultraviolet/violet photon and the *cis* configuration of the second C=C is changed into *trans*.

REFERENCES

[1] Nathan CF. Respiratory burst in adherent human neutrophils: triggering by colony-stimulating factors CSF-GM and CSF-G. *Blood.* 1989; 73: 301-6.

[2] Nathan CF. Neutrophil activation on biological surfaces. Massive secretion of hydrogen peroxide in response to products of macrophages and lymphocytes. *J Clin Invest.* 1987; 80: 1550-60.

[3] Stief TW. The blood fibrinolysis / deep-sea analogy: a hypothesis on the cell signals singlet oxygen/photons as natural antithrombotics. *Thromb Res.* 2000; 99: 1-20.

[4] Stief TW. Neutrophil Granulocytes in Hemostasis. In: *Handbook of Hematology Research: Hemorheology, Hemophilia and Blood Coagulation*; Tondre R, Lebègue C, eds; Nova Science Publishers; New York; 2011; pp. 179-200.

[5] Stief TW. The physiology and pharmacology of singlet oxygen. *Med Hypoth.* 2003; 60: 567-572.

[6] Stief TW. Regulation of hemostasis by singlet-oxygen ($^1\Delta O_2$). *Curr Vasc Pharmacol.* 2004; 2: 357-362.

[7] Stief TW. Hemostasis tolerable singlet oxygen - a perspective in AIDS therapy. *Hemost Lab* 2008; 1: 21-40.

[8] Stief T. Light quants of low wave length prime blood neutrophils for ROS generation. *Hemost Lab* 2013; 6 (issue 4)

[9] Sabbah S, Hui J, Hauser FE, Nelson WA, Hawryshyn CW. Ontogeny in the visual system of Nile tilapia. *J Exp Biol.* 2012; 215 (Pt 15): 2684-95.

[10] Davies WI, Zheng L, Hughes S, Tamai TK, Turton M, Halford S, Foster RG, Whitmore D, Hankins MW. Functional diversity of melanopsins and their global expression in the teleost retina. *Cell Mol Life Sci.* 2011; 68: 4115-32.

[11] Nava SS, An S, Hamil T. Visual detection of UV cues by adult zebrafish (Danio rerio). *J Vis.* 2011; 11: 2.

[12] Watanabe K, Nishimura Y, Oka T, Nomoto T, Kon T, Shintou T, Hirano M, Shimada Y, Umemoto N, Kuroyanagi J, Wang Z, Zhang Z, Nishimura N, Miyazaki T, Imamura T, Tanaka T. In vivo imaging of zebrafish retinal cells using fluorescent coumarin derivatives. *BMC Neurosci.* 2010; 11: 116.

[13] Hawryshyn CW. Ultraviolet polarization vision and visually guided behavior in fishes. *Brain Behav Evol*. 2010; 75: 186-94.

[14] Gan KJ, Novales Flamarique I. Thyroid hormone accelerates opsin expression during early photoreceptor differentiation and induces opsin switching in differentiated TRα-expressing cones of the salmonid retina. *Dev Dyn*. 2010; 239: 2700-13.

[15] Takechi M, Seno S, Kawamura S. Identification of cis-acting elements repressing blue opsin expression in zebrafish UV cones and pineal cells. *J Biol Chem*. 2008; 283: 31625-32.

[16] Cheng CL, Flamarique IN. Chromatic organization of cone photoreceptors in the retina of rainbow trout: single cones irreversibly switch from UV (SWS1) to blue (SWS2) light sensitive opsin during natural development. *J Exp Biol*. 2007; 210 (Pt 23): 4123-35.

[17] Kawano-Yamashita E, Terakita A, Koyanagi M, Shichida Y, Oishi T, Tamotsu S. Immunohistochemical characterization of a parapinopsin-containing photoreceptor cell involved in the ultraviolet/green discrimination in the pineal organ of the river lamprey *Lethenteron japonicum*. *J Exp Biol*. 2007; 210 (Pt 21): 3821-9.

[18] Cheng CL, Flamarique IN. Photoreceptor distribution in the retina of adult Pacific salmon: corner cones express blue opsin. *Vis Neurosci*. 2007; 24: 269-76.

[19] Cheng CL, Gan KJ, Flamarique IN. The ultraviolet opsin is the first opsin expressed during retinal development of salmonid fishes. *Invest Ophthalmol Vis Sci*. 2007; 48: 866-73.

[20] Bowmaker JK, Hunt DM. Evolution of vertebrate visual pigments. *Curr Biol*. 2006; 16: R484-9.

[21] Allison WT, Haimberger TJ, Hawryshyn CW, Temple SE. Visual pigment composition in zebrafish: Evidence for a rhodopsin-porphyropsin interchange system. *Vis Neurosci*. 2004; 21: 945-52.

[22] Cheng CL, Novales Flamarique I. Opsin expression: new mechanism for modulating colour vision. *Nature*. 2004; 428: 279.

[23] Takechi M, Hamaoka T, Kawamura S. Fluorescence visualization of ultraviolet-sensitive cone photoreceptor development in living zebrafish. *FEBS Lett*. 2003; 553: 90-4.

[24] Forsell J, Holmqvist B, Ekström P. Molecular identification and developmental expression of UV and greenopsin mRNAs in the pineal organ of the Atlantic halibut. *Brain Res Dev Brain Res*. 2002; 136: 51-62.

[25] Helvik JV, Drivenes O, Naess TH, Fjose A, Seo HC. Molecular cloning and characterization of five opsin genes from the marine flatfish Atlantic halibut (*Hippoglossus hippoglossus*). *Vis Neurosci.* 2001; 18: 767-80.

[26] Forsell J, Ekström P, Flamarique IN, Holmqvist B. Expression of pineal ultraviolet- and green-like opsins in the pineal organ and retina of teleosts. *J Exp Biol.* 2001; 204(Pt 14): 2517-25.

[27] Palacios AG, Goldsmith TH, Bernard GD. Sensitivity of cones from a cyprinid fish (*Danio aequipinnatus*) toultraviolet and visible light. *Vis Neurosci.* 1996; 13: 411-21.

[28] Raymond PA, Barthel LK, Stenkamp DL. The zebrafish ultraviolet cone opsin reported previously is expressed in rods. *Invest Ophthalmol Vis Sci.* 1996; 37: 948-50.

[29] Schmitt EA, Fadool JM, Dowling JE. Zebrafish ultraviolet cone opsin. *Invest Ophthalmol Vis Sci.* 1996; 37: 695.

[30] Goldsmith TH. Ultraviolet receptors and color vision: evolutionary implications and a dissonance of paradigms. *Vision Res.* 1994; 34: 1479-87.

[31] Raymond PA, Barthel LK, Rounsifer ME, Sullivan SA, Knight JK. Expression of rod and cone visual pigments in goldfish and zebrafish: a rhodopsin-like gene is expressed in cones. *Neuron.* 1993; 10: 1161-74.

[32] Palacios AG, Srivastava R, Goldsmith TH. Spectral and polarization sensitivity of photocurrents of amphibian rods in the visible and ultraviolet. *Vis Neurosci.* 1998; 15: 319-31.

[33] Stief T. The routine blood ROS generation assay (BRGA) triggered by typical septic concentrations of zymosan A. *Hemostasis Laboratory* 2013; 6: 89-98.

[34] Stief TW, Max M. Active endotoxin in sepsis. *Hemostasis Laboratory* 2008; 1: 53-60.

[35] Stief TW. Fungal sepsis can strongly increase plasmatic procalcitonin concentration. *Hemost Lab* 2012; 5: 27-33.

[36] Stief TW. Rapid diagnosis of fungal sepsis by the oxidative Limulus test. *Hemost Lab* 2010; 3: 289-96.

[37] Stief T. Micro-thrombi stimulate blood ROS generation. *Hemost Lab.* 2013; 6 (issue 2-3)

[38] McCaughan JS Jr. Photodynamic therapy: a review. *Drugs Aging.* 1999; 15: 49-68.

[39] Peplow PV, Chung TY, Ryan B, Baxter GD. Laser photobiomodulation of gene expression and release of growth factors and cytokines from

cells in culture: a review of human and animal studies. *Photomed Laser Surg.* 2011; 29: 285-304.

[40] Van Kets V, Karsten A, Davids LM. Laser light activation of a second-generation photosensitiser and its use as a potential photomodulatory agent in skin rejuvenation. *Lasers Med Sci.* 2012; 28: 589-95.

[41] Kubin A, Alth G, Jindra R, Jessner G, Ebermann R. Wavelength-dependent photoresponse of biological and aqueous model systems using the photodynamic plant pigment hypericin. *J Photochem Photobiol B.* 1996; 36: 103-8.

[42] Dobson R, Giovannoni G, Ramagopalan S. The month of birth effect in multiple sclerosis: systematic review, meta-analysis and effect of latitude. *J Neurol Neurosurg Psychiatry.* 2013; 84: 427-32.

[43] Meszaros ZS, Perl A, Faraone SV. Psychiatric symptoms in systemic lupus erythematosus: a systematic review. *J Clin Psychiatry.* 2012; 73: 993-1001.

[44] Sklar LR, Almutawa F, Lim HW, Hamzavi I. Effects of ultraviolet radiation, visible light, and infrared radiation on erythema and pigmentation: a review. *Photochem Photobiol Sci.* 2013 13; 12: 54-64.

[45] Hajrasouliha AR, Kaplan HJ. Light and ocular immunity. *Curr Opin Allergy Clin Immunol.* 2012 ; 12: 504-9.

[46] Guffey JS, Wilborn J. In vitro bactericidal effects of 405-nm and 470-nm blue light. *Photomed Laser Surg.* 2006; 24: 684-8.

[47] Dai T, Gupta A, Huang YY, Yin R, Murray CK, Vrahas MS, Sherwood M, Tegos GP, Hamblin MR. Blue light rescues mice from potentially fatal Pseudomonas aeruginosa burn infection: efficacy, safety, and mechanism of action. *Antimicrob Agents Chemother.* 2013; 57: 1238-45.

[48] Sawane M, Kajiya K. Ultraviolet light-induced changes of lymphatic and blood vasculature in skin and their molecular mechanisms. *Exp Dermatol.* 2012; 21 Suppl 1: 22-5.

[49] Reichrath J, Reichrath S. Hope and challenge: the importance of ultraviolet (UV) radiation for cutaneous vitamin D synthesis and skin cancer. *Scand J Clin Lab Invest Suppl.* 2012; 243: 112-9.

[50] Hockberger PE. A history of ultraviolet photobiology for humans, animals and microorganisms. *Photochem. Photobiol.* 2002; 76: 561–79.

[51] Hanson KM, Gratton E, Bardeen CJ. Sunscreen enhancement of UV-induced reactive oxygen species in the skin. *Free Radical Biology and Medicine* 2006; 41: 1205–12.

[52] Nair R, Maseeh A. Vitamin D: The "sunshine" vitamin. *J Pharmacol Pharmacother*. 2012; 3: 118-26.

[53] Zandi S, Kalia S, Lui H. UVA1 phototherapy: a concise and practical review. *Skin Therapy Lett*. 2012; 17: 1-4.

[54] Dai T, Vrahas MS, Murray CK, Hamblin MR. Ultraviolet C irradiation: an alternative antimicrobial approach to localized infections? *Exp Rev Anti Infect Ther*. 2012; 10: 185-95.

[55] www.wikipedia.org

[56] Sitrin RG, Pan PM, Harper HA, Todd RF 3rd, Harsh DM, Blackwood RA. Clustering of urokinase receptors (uPAR; CD87) induces proinflammatory signalling in human polymorphonuclear neutrophils. *J Immunol*. 2000; 165: 3341-9.

[57] Syrovets T, Lunov O, Simmet T. Plasmin as a proinflammatory cell activator. *J Leukoc Biol*. 2012; 2: 509-19.

[58] Cao D, Mizukami IF, Garni-Wagner BA, Kindzelskii AL, Todd RF 3rd, Boxer LA, Petty HR. Human urokinase-type plasminogen activator primes neutrophils for superoxide anion release. Possible roles of complement receptor type 3 and calcium. *J Immunol*. 1995; 154: 1817-29.

[59] Kindzelskii AL, Laska ZO, Todd RF 3rd, Petty HR. Urokinase-type plasminogen activator receptor reversibly dissociates from complement receptor type 3 (CD11b/CD18) during neutrophil polarization. *J Immunol*. 1996; 156: 297-309.

[60] Nygren H, Eriksson C, Lausmaa J. Adhesion and activation of platelets and polymorphonuclear granulocyte cells at TiO_2 surfaces. *J Lab Clin Med*. 1997; 129: 35-46.

[61] Bae YB, Oh H, Rhee SG, Yoo YD. Regulation of reactive oxygen species generation in cell signalling. *Mol. Cells*. 2011; 32: 491-509.

[62] Klotz LO, Kröncke KD, Sies H. Singlet oxygen-induced signaling effects in mammalian cells. *Photochem Photobiol Sci*. 2003; 2: 88-94.

[63] Stief TW. Nonradical excited oxygen species induce selective thrombolysis in vivo. *Thromb Res*. 1991; 62: 147-63.

[64] Stief T. Cortisol´s suppression of blood ROS generation quantified. *Hemostasis Laboratory* 2013; 6 (issue 2-3)

[65] Del Maestro R, Thaw HH, Björk J, Planker M, Arfors KE. Free radicals as mediators of tissue injury. *Acta Physiol Scand Suppl*. 1980; 492: 43-57.

[66] Sluiter W, de Vree WJ, Pietersma A, Koster JF. Prevention of late lumen loss after coronary angioplasty by photodynamic therapy: role of activated neutrophils. *Mol Cell Biochem*. 1996; 157: 233-8.

[67] Plaetzer K, Krammer B, Berlanda J, Berr F, Kiesslich T. Photophysics and photochemistry of photodynamic therapy: fundamental aspects. *Lasers Med Sci*. 2009; 24: 259-68.

[68] Verhille M, Couleaud P, Vanderesse R, Brault D, Barberi-Heyob M, Froschot C. Modulation of photosensitization processes for an improved targeted photodynamic therapy. *Curr Med Chem*. 2010; 17: 3925-43.

[69] Stief TW, Fareed J. The antithrombotic factor singlet oxygen/light ($^1O_2$/hν). *Clin Appl Thrombosis/Hemostasis* 2000; 6: 22-30.

[70] Stief TW, Feek U, Ramaswamy A, Kretschmer V, Renz H, Fareed J. Singlet oxygen ($^1O_2$) disrupts platelet aggregates. *Thromb Res*. 2001; 104/5: 361-369.

[71] Czesnikiewicz-Guzik M, Lorkowska B, Zapala J, Czajka M, Szuta M, Loster B, Guzik TJ, Korbut R. NADPH oxidase and uncoupled nitric oxide synthase are major sources of reactive oxygen species in oral squamous cell carcinoma. Potential implications for immune regulation in high oxidative stress conditions. *J Physiol Pharmacol*. 2008; 59: 139-52.

Chapter 3

Blood Neutrophils Alert Each Other by Photons

Abstract

Background: The neutrophils are the strongest cells of innate immunology in human blood. They destroy dangerous pathogens such as bacteria, fungi, parasites, or micro-thrombi. Their main weapons are NADPH-oxidase/myeloperoxidase that generate large amounts of hydrogen peroxide (H_2O_2) and hypochlorite (HOCl). In critical situations they recruit other blood neutrophils. The mechanisms of this self-amplification of neutrophils are poorly understood. Neutrophils generate and perceive photons. Therefore, photons could be important in inter-cellular communication in the darkness of the blood vessel.

Material and Methods: 150 µl Hanks′ Balanced Salt Solution (HBSS) were incubated in transparent polystyrene Brand®781602 F-wells with 10 µl normal citrated blood, 10 µl 5 mM luminol in 0.9% NaCl, and 3.8 µg/ml (final) zymosan A (ZyA). The blood ROS generations in these wells and in the 0 µg/ml ZyA containing neighbor wells (NW) were determined by a photons-multiplying microtiter plate photometer (LUmo). The proportion of photons that passed from the illuminating well to its NW was calculated. At 156 min pre-incubation time in absence of ZyA 10 µl 36 µg/ml ZyA (2 µg/ml final) were added in 4-fold to wells with illuminating NW or to wells without illuminating NW. Furthermore, 125 µl HBSS were incubated with 10 µl normal citrated blood, 0 or 0.3 mM luminol, and 2.2 µg/ml (final) ZyA. After 62 min the NW of illuminating or not illuminating wells were assayed for luminol-enhanced light emission, triggered by 0.6 µg/ml ZyA.

Results and Discussion: Blood ROS generation was about 2-fold enhanced if the NW was illuminating, independent if the main light of the

NW was luminol derived blue or cell derived UV to violet. In the initial phase of blood ROS generation about 20% of the photons passed to the NW. At begin of cellular fibrinolysis, less than 10% of the photons passed to the NW. Finally, only 5% of the photons passed to the NW. The neutrophils seem to preferably generate UV-photons if they detect dangerous enemies, such as micro-clots. UV-photons seem to be long-lasting alert signals; violet or blue photons could be signals of a lower alert category.

Keywords: Neutrophils, reactive oxygen species, ROS, singlet oxygen, photons, blue, violet, opsin

INTRODUCTION

The polymorphonuclear neutrophil granulocytes (neutrophils) are the main cells of innate immunology in human blood. They destroy life threatening pathogens such as bacteria, fungi, parasites, or micro-thrombi [1-4].

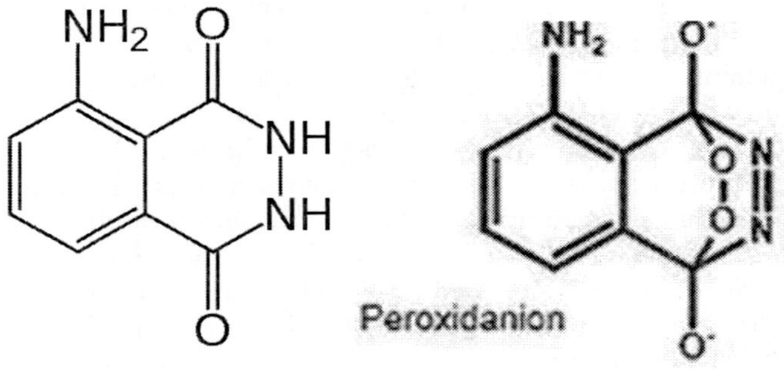

Figure 1. Chemical structure of luminol (left) and of its unstable peroxide (right) [17]. 5-Amino-2,3-dihydro-1,4-phthalazine-dione ($C_8H_7N_3O_2$; MW: 177.17 Daltons) is oxidized e.g. by H_2O_2 or $^1O_2^*$ to its unstable peroxide that spontaneously decays releasing a blue photon.

The principal weapon of activated neutrophils is the combination of membranous NADPH-oxidase (Nox-2) and secreted myeloperoxidase that together generate large amounts of hydrogen peroxide (H_2O_2), the ROS mother and her two daughters hydroxyl-radical (•OH) and non-radicalic excited singlet molecular oxygen ($^1O_2^*$) [5-12]. $^1O_2^*$ reacts with C=C to excited carbonyls (R-C=O*) that emit light in the violet/blue spectrum.

In critical situations the neutrophils call other blood neutrophils to amplify inflammation [13,14]. The important mechanisms of auto-recruitment of neutrophils are poorly understood. Neutrophils generate and perceive photons [15,16]. Therefore, photons in the blue spectrum around 400 nm (Figure 1) or in the UV spectrum could be of great importance in inter-cellular communication in the darkness of the blood stream [14].

MATERIAL AND METHODS

Blood ROS generation was measured by the BRGA (blood ROS generation assay) [18]. 150 µl Hanks′ Balanced Salt Solution (HBSS; Sigma, Deisenhofen, Germany) were incubated (4-fold determinations, intra-assay CV < 10%) in transparent polystyrene flat-bottomed wells (Brand, Wertheim, Germany; article nr. 781602; F-wells) with 10 µl normal citrated blood (venous blood drawn after written informed consent in newest polypropylene monovettes of Sarstedt, Nümbrecht, Germany with 11 mM Na_3-citrate final blood conc.; stored for 1d 23°C;) 10 µl 5 mM luminol sodium salt (Sigma) in 0.9% NaCl, and 3.8 µg/ml (final) zymosan A (ZyA; Sigma) in 0.9% NaCl. The blood ROS generations in these wells and in 0 µg/ml ZyA containing neighbor wells (NW) were determined by a photons-multiplying microtiter plate photometer (LUmo; Autobio Labtec - anthos, Krefeld, Germany). The proportion of photons that passed from the illuminating well to its NW was calculated. At 156 min pre-incubation time in absence of ZyA 10 µl 36 µg/ml ZyA (2 µg/ml final) were added in 4-fold to wells with illuminating NW or to wells without illuminating NW. The specific light count per well was calculated correcting for passed photons out of the NW.

Furthermore, 125 µl HBSS were incubated with 10 µl normal citrated blood, 0 mM or 0.3 mM (final) luminol, and 2.2 µg/ml (final) ZyA. After 62 min the NW of illuminating or not illuminating wells were assayed for luminol-enhanced light emission, triggered by 0.6 µg/ml ZyA.

HBSS was 185.4 mg/l $CaCl_2 \cdot 2 H_2O$, 200 mg/l $MgSO_4 \cdot 7 H_2O$, 400 mg/l KCl, 60 mg/l KH_2PO_4, 350 mg/l $NaHCO_3$, 8000 mg/l NaCl, 90 mg/l Na_2HPO_4, 1000 mg/l glucose, pH 7.0-7.4. Expressed in molarity, the concentrations of the HBSS components are: 1.3 mM Ca^{2+}, 0.8 mM Mg^{2+}, 5.8 mM K^+, 143 mM Na^+, 144 mM Cl^-, 1.6 mM SO_4^{2-}, 0.4 mM $H_2PO_4^-$, 0.6 mM HPO_4^{2-}, 4.2 mM HCO_3^-, 5.6 mM glucose.

Results and Discussion

At about 1h reaction time the maximal ROS generation of about 4500 RLU/s triggered by 3.8 µg/ml ZyA was reached. Then the blood ROS generation decreased to a value of about 1800 RLU/s at 156 min, followed by a new increase in blood ROS generation – typical for the onset of cellular fibrinolysis [19] - with a second maximum of about 3700 RLU/s, reached at about 200 min. At 272 min the maximum was left, the blood ROS generation decreased to about 1300 RLU/s at 326 min (Figure 2a). Figure 2b demonstrates the proportion of photons detected in the neighbor well (NW) of a luminol-blue illuminating well. At begin of blood ROS generation about 20% of the photons passed to the NW. At 149 min still 16% of the photons passed to the NW. But at more than 156 min incubation time, at pronounced cellular fibrinolysis, less than 10% of the photons passed to the NW. Finally, only 5% of the photons passed to the NW. This means that in the phase of the second maximum, the cellular fibrinolysis phase, much more UV-photons were generated. The neutrophils seem to preferably generate UV-photons if they detect a dangerous enemy, such as a micro-clot. UV-photons seem to be a signal of extreme alert, violet or blue photons seem to be signals of a lower alert category.

The blood ROS generation was about 2-fold enhanced if the NW was illuminating (Figure 3).

With the same blood samples being stored for another day at 23°C, that pre-activates AM-coagulation (altered matrix coagulation), the blood ROS generation kinetic changed towards accelerated activation and accelerated begin of cellular fibrinolysis at about 1h (Figure 4). Blood should be added to HBSS and not directly to the polystyrene plate that activates the neutrophils.

Figure 5 demonstrates the blood ROS generation in 1h luminol/ZyA-treated wells, figure 6 the blood ROS generation in 1h ZyA-treated wells. This means that the illumination of the wells of figure 5 is primarily due to luminol-blue, whereas the illumination of the wells of figure 6 is primarily due to the own photons of activated neutrophils (violet to UV photons).

With luminol-blue light in the NW the blood ROS generations in the illuminated wells increased up to 4.5-fold at 20 min incubation time, the usual increase factor was about 2. With UV to violet light in the NW the blood ROS generations in the illuminated wells increased up to 2-fold at 270 min incubation time, i.e. UV to violet photons have a long-lasting stimulating action on blood ROS generation (Figures 7-9).

There are really exciting times ahead of all researchers who focus on the biochemistry, pathobiochemistry, physiology, pathophysiology, pharmacology, toxicology, histology, histopathology of singlet oxygen [20-23].

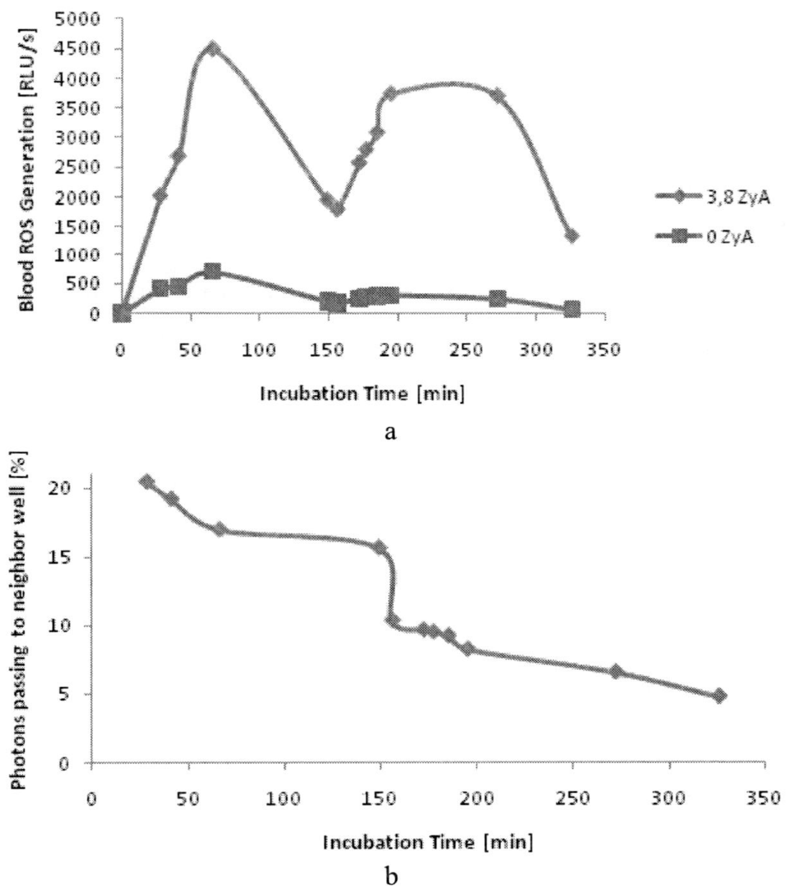

Figure 2. Kinetic of blood ROS generation. 150 µl HBSS were incubated in 8-fold in transparent polystyrene Brand®781602 F-wells with 10 µl normal citrated blood (stored for 1d 23°C), 10 µl 5 mM luminol in 0.9% NaCl, and 0 or 3.8 µg/ml (final) zymosan A (ZyA). The blood ROS generations in these wells and in the 0 µg/ml ZyA containing neighbor wells (NW) were determined by a photons-multiplying microtiter plate photometer (LUmo). 0 µg/ml ZyA control wells without illuminating NW always had 0 relative light units per second (RLU/s) (Figure 2a). The proportion of photons that passed from the illuminating well to its NW was calculated (Figure 2b).

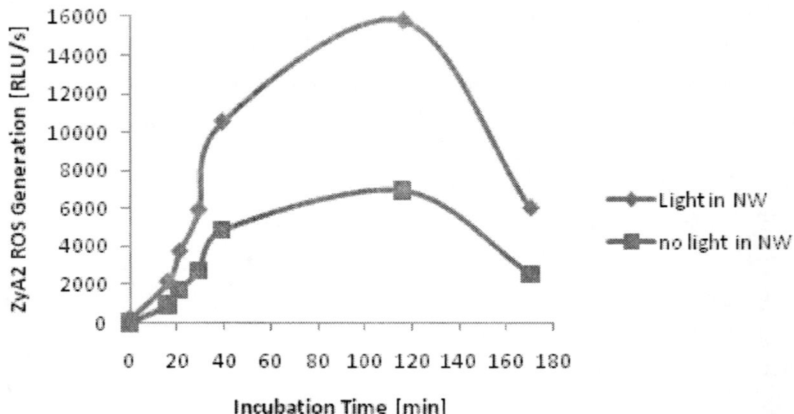

Figure 3. Kinetic of blood ROS generation in neighbor wells after 156 min pre-incubation. The experiment was performed as described in figure 1. At 156 min pre-incubation time in absence of ZyA 10 µl 36 µg/ml ZyA (2 µg/ml final = ZyA2) were added in 4-fold to wells with light in their neighbor wells (NW) or to wells without light in their NW. The blood ROS generations were determined by a photons-multiplying microtiter plate photometer (LUmo).

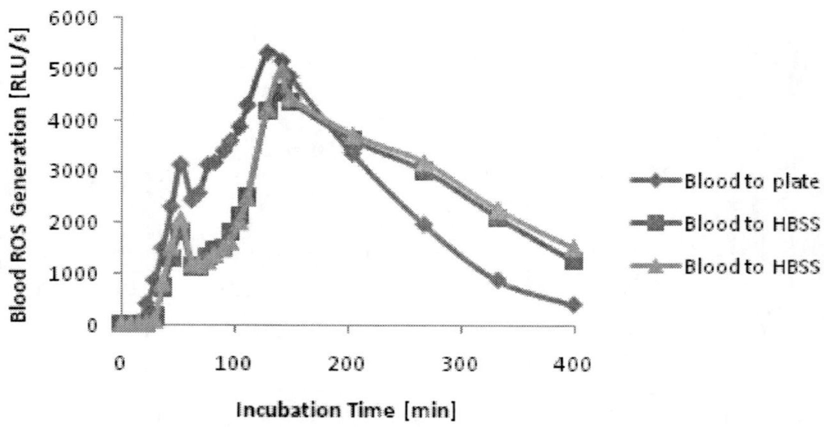

Figure 4. Kinetic of blood ROS generation. 125 µl HBSS were incubated in 2-fold (■▲) in transparent polystyrene Brand®781602 F-wells with 10 µl normal citrated blood (stored for 2d 23°C), 0.3 mM luminol in 0.9% NaCl, and 2.2 µg/ml (final) zymosan A (ZyA). The blood ROS generations were determined by a photons-multiplying microtiter plate photometer (LUmo). In an alternative approach (♦) the 10 µl blood samples were first added to the polystyrene plate and then HBSS was added.

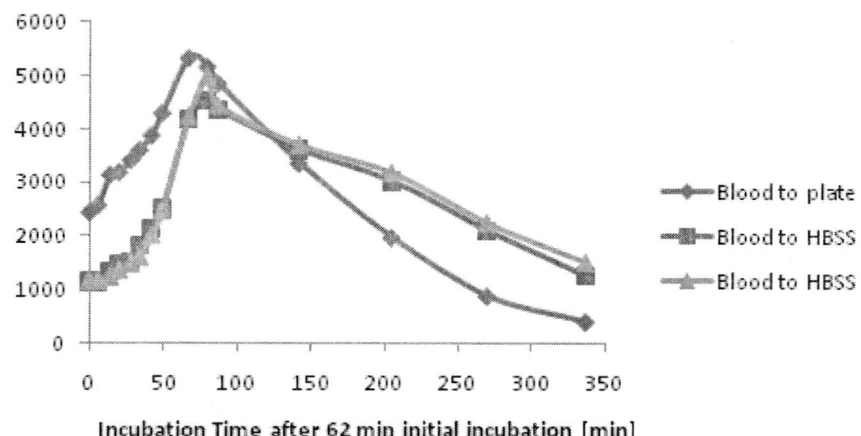

Figure 5. Final incubation for blood ROS determination. Figure 4 is reproduced with 62 min initial incubation time point being the new 0 point.

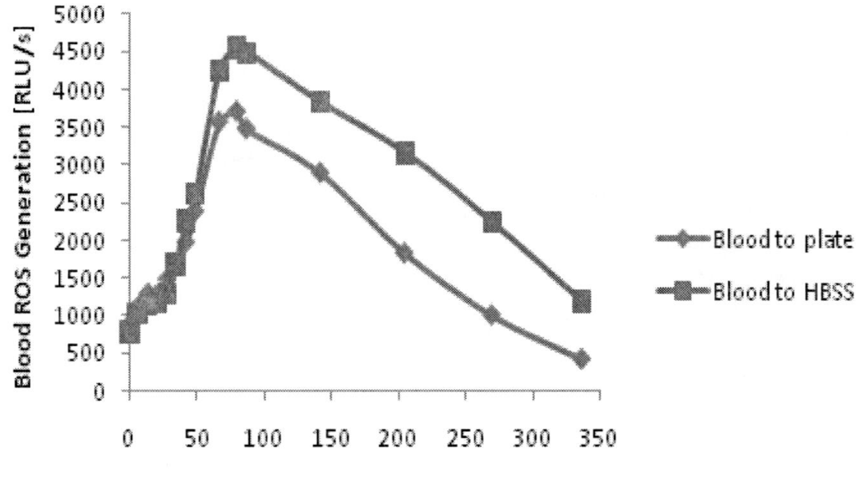

Figure 6. Final incubation for blood ROS determination in luminol absence but ZyA presence within the initial 62 min incubation. The experiment of figure 4 was performed in absence of luminol within the initial 62 min incubation. Then 0.3 mM luminol and 0.6 μg/ml ZyA were added at the new 0 min time point.

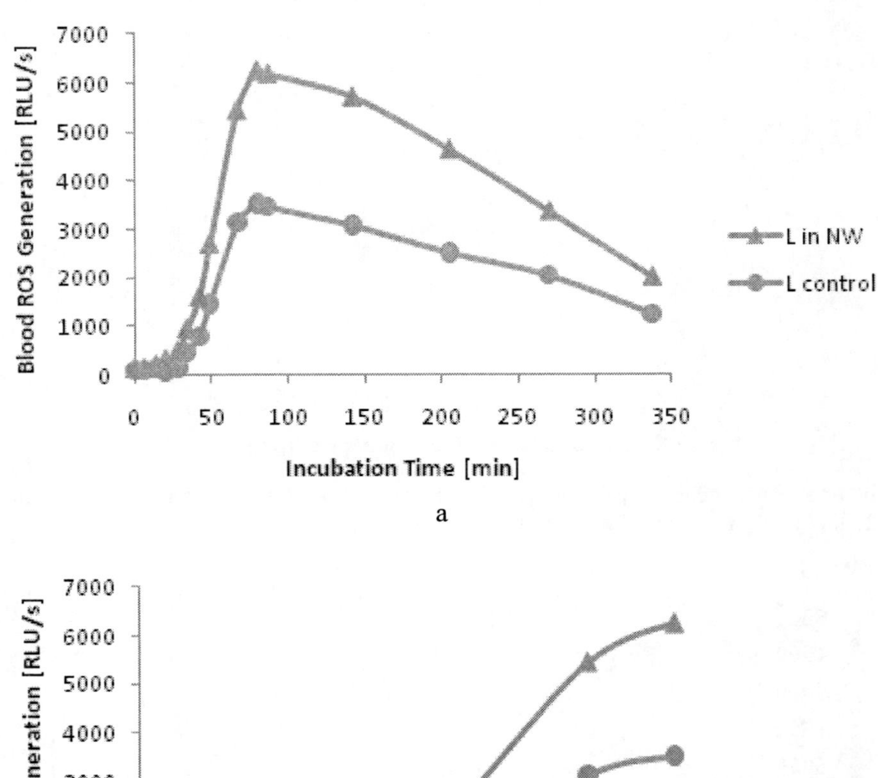

Figure 7. Blood ROS generation in neighbor wells of luminol/ZyA treated wells. The blood ROS generations in neighbor wells (NW) of 0.3 mM luminol (L) + 2.2 µg/ml ZyA treated wells (or untreated control wells) were determined, adding at the 62 min time point = new 0 min point 0.3 mM luminol + 0.6 µg/ml ZyA to the reaction wells.

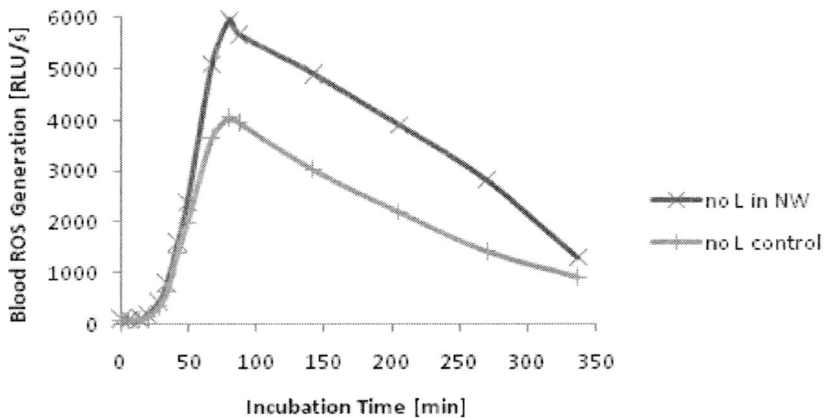

Figure 8. Blood ROS generation in neighbor wells of ZyA treated wells. The blood ROS generations in neighbor wells of 2.2 µg/ml ZyA (without luminol) treated wells (or untreated control wells) were determined in another sample, adding at the 62 min time point = new 0 min point 0.3 mM luminol + 0.6 µg/ml ZyA to the reaction wells and measuring the light emission in a photons-multiplyer microtiter plate luminometer (LUmo).

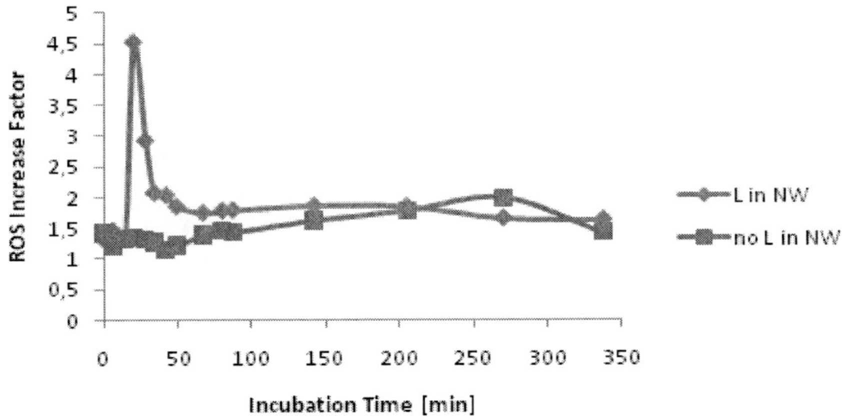

Figure 9. ROS increase in NW of luminol/ZyA treated wells (L) or in NW of ZyA treated wells (no L). The ratio between blood ROS generation in NW of L-wells divided by L-control wells and the ratio between blood ROS generation in NW of no L-wells divided by no L-control wells was calculated.

References

[1] Stief TW. Neutrophil granulocytes in hemostasis. *Hemostasis Laboratory* 2009; 2: 269-89.

[2] Alves-Filho JC, de Freitas A, Spiller F, Souto FO, Cunha FQ. The role of neutrophils in severe sepsis. *Shock.* 2008; 30 Suppl 1: 3-9.

[3] Goodridge HS, Wolf AJ, Underhill DM. Beta-glucan recognition by the innate immune system. *Immunol Rev*. 2009; 230: 38-50.

[4] Joos C, Marrama L, Polson HE, Corre S, Diatta AM, Diouf B, Trape JF, Tall A, Longacre S, Perraut R. Clinical protection from falciparum malaria correlates with neutrophil respiratory bursts induced by merozoites opsonized with human serum antibodies. *PLoS One*. 2010; 5: e9871.

[5] Rosen H, Klebanoff SJ. Formation of singlet oxygen by the myeloperoxidase-mediated antimicrobial system. *J Biol Chem*. 1977; 252: 4803-10.

[6] Nathan CF. Neutrophil activation on biological surfaces. Massive secretion of hydrogen peroxide in response to products of macrophages and lymphocytes. *J Clin Invest*. 1987; 80: 1550-60.

[7] Briviba K, Klotz LO, Sies H. Toxic and signaling effects of photochemically or chemically generated singlet oxygen in biological systems. *Biol Chem*. 1997; 378: 1259-65.

[8] Kiryu C, Makiuchi M, Miyazaki J, Fujinaga T, Kakinuma K. Physiological production of singlet molecular oxygen in the myeloperoxidase-H_2O_2-chloride system. *FEBS Lett*. 1999; 443: 154-8.

[9] Seguchi H, Kobayashi T. Study of NADPH oxidase-activated sites in human neutrophils. *J Electron Microsc*. 2002; 51: 87-91.

[10] Stief TW. The physiology and pharmacology of singlet oxygen. *Med Hypoth.* 2003; 60: 567-72.

[11] Stief TW. Regulation of hemostasis by singlet-oxygen ($^1\Delta O_2$). *Curr Vasc Pharmacol* 2004; 2: 357-62.

[12] Maghzal G, Krause KH, Stocker R, Jaquet V. Detection of reactive oxygen species derived from the family of NOX (NADPH oxidases). *Free Radic Biol Med*. 2012; 53: 1903-18.

[13] Stief TW, Fareed J. The antithrombotic factor singlet oxygen/light ($^1O_2/h\nu$). *Clin Appl Thrombosis/Hemostasis* 2000; 6: 22-30.

[14] Stief TW. The blood fibrinolysis / deep - sea analogy: a hypothesis on the cell signals singlet oxygen / photons as natural antithrombotics. *Thromb Res*. 2000; 99: 1-20.

[15] Stief T. Light quants of low wave length prime blood neutrophils for ROS generation. *Hemost Lab* 2013; 6 (issue 4)

[16] Stief T. Blood neutrophils see UV light: 340 nm ultraviolet A stimulates blood ROS generation nearly half as strong as 405 nm violet photons. *Hemostasis Laboratory* 2013; 6 (issue 4).

[17] www.wikipedia.org or www.wikipedia.de

[18] Stief T. The routine blood ROS generation assay (BRGA) triggered by typical septic concentrations of zymosan A. *Hemost Lab*. 2013; 6: 89-98.

[19] Stief T. Micro-thrombi stimulate blood ROS generation. *Hemostasis Laboratory* 2013; 6 (issue 2-3)

[20] Ano Y, Sakudo A, Onodera T. Role of microglia in oxidative toxicity associated with encephalomyocarditis virus infection in the central nervous system. *Int J Mol Sci*. 2012; 13: 7365-74.

[21] Savina A, Vargas P, Guermonprez P, Lennon AM, Amigorena S. Measuring pH, ROS production, maturation, and degradation in dendritic cell phagosomes using cytofluorometry-based assays. *Methods Mol Biol*. 2010; 595: 383-402.

[22] Legrand-Poels S, Schoonbroodt S, Matroule JY, Piette J. Nf-kappa B: an important transcription factor in photobiology. *J Photochem Photobiol B*. 1998; 45: 1-8.

[23] Fortin CF, McDonald PP, Lesur O, Fülöp T Jr. Aging and neutrophils: there is still much to do. *Rejuvenation Res*. 2008; 11: 873-82.

Chapter 4

Singlet Oxygen ($^1O_2^*$) Primes Blood Neutrophils to Generate ROS

Abstract

Background: Excited singlet molecular oxygen ($^1O_2^*$) is both a defensive weapon and a cell signal. In high concentrations $^1O_2^*$ destroys fungi, bacteria, parasites, or thrombi. In low concentrations $^1O_2^*$ is an inter- or intra-cellular communication signal. The main blood cells that use $^1O_2^*$ for destruction and signaling are the neutrophils. Neutrophils′ NADPH-oxidase assembly for blood ROS generation is stimulated by photons with wave lengths in the range around 300-400 nm. These photons stem from excited oxygen, the oxygen being bound in an excited carbonyl (R-C=O*) that has been formed by reaction of $^1O_2^*$ with R_1-C=C-R_2.

Material and Methods: 125 µl Hanks′ Balanced Salt Solution were incubated in triplicate with 10 µl freshest normal citrated blood or EDTA-blood, 10 µl 5 mM (final: 0.3 mM) luminol, and 5 µl 18 µg/ml (final: 0.5-0.6 µg/ml) zymosan A (ZyA) in black high quality polystyrene microwells (Brand®781608). After 0-324 min (37°C) 40 µl 0.5 mM chloramine-T® (N-chloro-toluene-sulfonamide) in 0.9% NaCl, 2.5 mM Na_3-citrate, pH 7.4 were added. The plates were incubated at 37°C in a sterile incubator and the blood ROS generation assay (BRGA) was repetitively measured in a photons-multiplyer microtiter plate luminometer (LUmo) with an integration time of 0.5s. In an alternative approach the ZyA was added after 60 min (37°C).

Results: The normal blood ROS generations in 0.6 µg/ml ZyA-triggered freshest citrated samples ran with at least 3 maxima, the first at 60-90 min (about 50 RLU/s), the second (about 25 RLU/s) at 140-170 min, the third (about 11 RLU/s) at 240-260 min. The normal blood ROS generation in slightly triggered freshest EDTA samples also had at least 3 maxima, the first at 60-70 min (about 90 RLU/s), the second (about 50 RLU/s) at about 170 min, the third (about 25 RLU/s) at about 240 min. Addition of 25 nmoles of chloramine to the 10 µl citrated blood resulted in a 10-fold increased ROS generation maximum of 314 RLU/s at 181 min. In oxidized citrated blood from 218 min to at least 324 min the blood ROS generation maintained stable at an intensity of about 50% of maximum, whereas in unoxidized controls the blood ROS generation approached the zero-line. 0.5 mM chloramine-oxidized EDTA-blood resulted in 2-fold increased blood ROS generation at about 0.5h oxidation time and up to 7-fold blood ROS generation at about 3h oxidation time.

Discussion: 6.3 nmoles of CT given to 10 µl citrated blood means that chloramines in 0.5-1 mM concentrations in blood are strongly thrombolytic. Here 4-fold this amount of chloramine was added and more than 10-fold enhancement of blood ROS generation appeared, i.e. chloramine concentrations of about 2-4 mM that could well be reached in the micro-environment of activated neutrophils are strongly fibrinolytic. In normal blood about 50% of the NADPH-oxidase capacity is kept in reserve for following activations, if necessary, e.g. against life-threatening micro-thrombi. The regulation of blood ROS generation by $^1O_2^*$ is of great medical interest.

Keywords: Singlet oxygen, chloramine, reactive oxygen species, ROS, neutrophils, excited carbonyls, BRGA

INTRODUCTION

Singlet spin state (^{1}) di-oxygen (O_2) is a molecule in excited (*) form, i.e. $^1O_2^*$ releases a photon (hv) when it returns to normal ground state oxygen ($^1O_2^* \rightarrow O_2 + h\nu$). If the excited oxygen atom is bound in a carbonyl (R-C=O*) then the biologically so important light in the range around 300-400 nm is set free [1-4]. In low concentrations $^1O_2^*$ is an inter- or intra-cellular communication signal [5]. The main blood cells that use $^1O_2^*$ signaling (but also for destruction) are the neutrophils [6-12]. Neutrophils′ NADPH-oxidase assembly for blood ROS generation is stimulated by 300-400 nm photons [13-15]. The regulation of blood ROS generation by $^1O_2^*$ is of great medical interest [16,17].

MATERIAL AND METHODS

125 µl Hanks′ Balanced Salt Solution (HBSS) were incubated in triplicate with 10 µl freshest (< 0.5h old) normal citrated blood (11 mM Na_3-citrate final conc. in an at least still 1 year valid polypropylene monovette from Sarstedt, Nümbrecht, Germany) or 10 µl freshest EDTA-blood (1.6 mg/ml K_3-EDTA in 2.6 ml venous blood drawn after written informed consent into a Sarstedt - polypropylene monovette), 10 µl 5 mM (final: 0.3 mM) luminol (Sigma, Deisenhofen, Germany; article nr. A4685-1G; 1g dissolved in 25.1 ml 0.9 % NaCl = 200 mM stem solution), and 5 µl 18 µg/ml (final: 0.5-0.6 µg/ml) zymosan A (ZyA; Z-4250-1G, lot nr. 27H0495; Sigma) in black high quality flat-bottomed polystyrene microwells (Brand, Wertheim, Germany; article nr. 781608).

After 0-323 min (37°C) 40 µl 0.5 mM chloramine-T® (N-chloro-toluene-sulfonamide trihydrate; Sigma; MW: 282 Daltons) in 0.9% NaCl, 2.5 mM Na_3-citrate, pH 7.4 were added. The plates were incubated at 37°C in a sterile incubator and the BRGA was repetitively measured in a photons-multiplyer microtiter plate luminometer (LUmo; anthos, Krefeld, Germany) with an integration time per well of 0.5s [17]. In an alternative approach (defined as BRGA-60-) the ZyA was added after a first incubation time of 60 min (37°C).

The intra-assay coefficients of variation were less than 15% in the BRGA and less than 10% in the BRGA-60-. Critical differences between main values and control values were tested in the $X^2\bar{x}$ test [18].

The LUmo analyzer can be calibrated by pure chloramine-T®/luminol. 10 µl 5 mM luminol in 0.9 % NaCl were incubated with 0-100 µl 0.5 mM chloramine-T® in 0.9% NaCl, 2.5 mM Na_3-citrate, pH 7.4 in black polystyrene wells. The light emission per well was measured by a LUmo.

HBSS (modified without phenol red; article nr. 55037C-1000ML) was: 185.4 mg/l $CaCl_2 \cdot 2 H_2O$, 200 mg/l $MgSO_4 \cdot 7 H_2O$, 400 mg/l KCl, 60 mg/l KH_2PO_4, 350 mg/l $NaHCO_3$, 8000 mg/l NaCl, 90 mg/l Na_2HPO_4, 1000 mg/l glucose, pH 7.0-7.4. In molarity, the conc. of the HBSS components are: 1.3 mM Ca^{2+}, 0.8 mM Mg^{2+}, 5.8 mM K^+, 143 mM Na^+, 144 mM Cl^-, 1.6 mM SO_4^{2-}, 0.4 mM $H_2PO_4^-$, 0.6 mM HPO_4^{2-}, 4.2 mM HCO_3^-, 5.6 mM glucose.

RESULTS AND DISCUSSION

Normal blood ROS generations in slightly triggered (0.6 µg/ml ZyA) freshest citrated samples ran with at least 3 maxima, the first at 60-90 min (about 50 RLU/s), the second (about 25 RLU/s) at 140-170 min, the third (about 11 RLU/s) at 240-260 min (Figure 1). Within several minutes the recalcified (citrated or EDTA) blood clotted to a micro-thrombus that stimulated further NADPH-oxidase assemblies [19,20]. There could be further maxima of about 5, 3, 2 etc. RLU/s detectable at repetitively two-fold the incubation time by future luminescence machines with increased sensitivity.

The normal blood ROS generation in slightly triggered (0.6 µg/ml ZyA) freshest EDTA samples also had at least 3 maxima, the first at 60-70 min (about 90 RLU/s), the second (about 50 RLU/s) at about 170 min, the third (about 25 RLU/s) at about 240 min (Figure 2). There could be further maxima of about 12, 6, 3, etc. RLU/s detectable at repetitively twofold the incubation time.

This means that the activated neutrophil always keeps 50% of its NADPH-oxidase capacity in reserve for future activations if required, e.g. against life-threatening micro-thrombi. The CD11b/18 induced triggering of the NADPH-oxidase [21] is favoured by EDTA, that tears off structural Ca^{2+} out of proteins, an action comparable to the inflammatory molecules of the S100 family (calprotectin, S100A8) [22,23].

Addition of 25 nmoles of chloramine to the 10 µl citrated blood resulted in a ROS generation maximum of 314 RLU/s at 181 min. Unoxidized controls had only 15 RLU/s at 181 min. In oxidized blood from 218 min to at least 324 min incubation the blood ROS generation maintained stable at an intensity of about 50% of maximum, whereas in unoxidized controls the blood ROS generation approached the zero-line (Figure 3). However, blood ROS generation in oxidized blood not started until 80 min whereas in unoxidized blood only about 20 min were needed to initiate blood ROS generation. Highest concentrations of singlet oxygen transiently damage the eyes of our neutrophils, however singlet oxygen primes the neutrophils for prolonged and increased NADPH-oxidase activity.

In the BRGA-60- version of the assay, i.e. addition of ZyA not until 60 min reaction time, the blood ROS generation increased to values higher than 600 RLU/s at the end of the incubation time (Figure 4).

The BRGA-60- version of the assay performed with normal EDTA-blood resulted in maximal blood ROS generations of about 240 RLU/s reached at about 300 min (Figure 5).

0.5 mM chloramine-oxidized EDTA-blood resulted in 2-fold increased blood ROS generation at about 0.5h oxidation time and up to 7-fold blood ROS generation at about 3h oxidation time (Figure 6).

100 µl 0.5 mM chloramine and 10 µl 5 mM luminol gave about 400 RLU/s light emission in the LUmo (Figure 7). The singlet oxygen generated by chloramine (Figure 8) reacts with luminol to luminol-peroxide that decays releasing a blue photon.

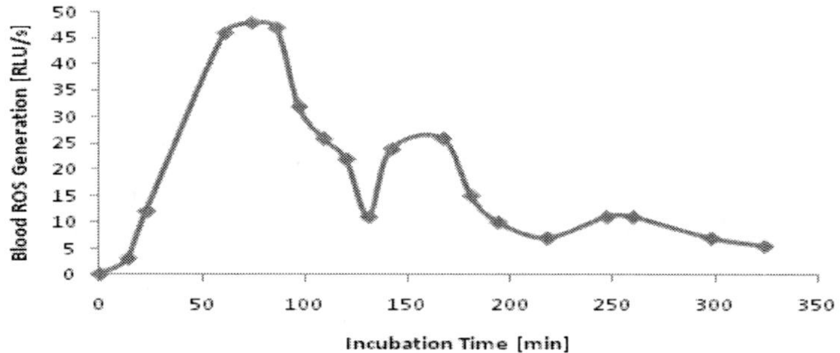

Figure 1. Kinetic of blood ROS generation in freshest normal citrated blood. Normal citrated blood (less than 0.5h old) was analyzed in the blood ROS generation assay (BRGA; 125 µl HBSS in black Brand®781608 wells, 10 µl blood, 10 µl 5 mM luminol, 5 µl 18 µg/ml zymosan A). After 0-324 min (37°C) the light emissions per well were measured by a photons-multiplying microtiter plate photometer (LUmo).

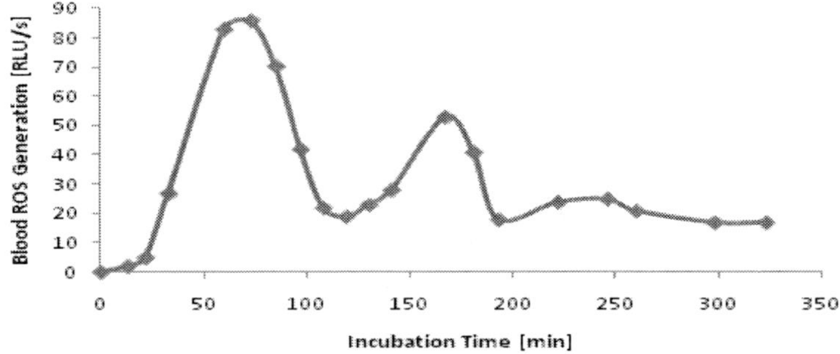

Figure 2. Kinetic of blood ROS generation in freshest normal EDTA-blood. Normal EDTA-blood (less than 0.5h old) was analyzed in the blood ROS generation assay (BRGA; 125 µl HBSS in black Brand®781608 wells, 10 µl blood, 10 µl 5 mM luminol, 5 µl 18 µg/ml zymosan A). After 0-323 min (37°C) the light emissions per well were measured by a photons-multiplying microtiter plate photometer (LUmo).

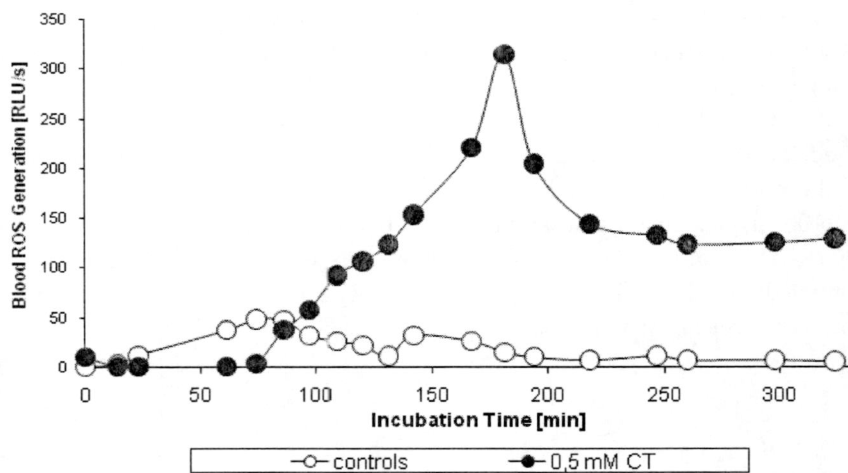

Figure 3. Kinetic of blood ROS generation in oxidized normal citrated blood. Normal citrated blood (less than 0.5h old) was analyzed in the blood ROS generation assay (BRGA; 125 µl HBSS in black Brand®781608 wells, 10 µl blood, 10 µl 5 mM luminol, 5 µl 18 µg/ml zymosan A) adding 40 µl 0.5 mM (25 nmoles) chloramine-T®. After 0-324 min (37°C) the light emissions per well were measured by a photons-multiplying microtiter plate photometer; controls: no oxidant added (O).

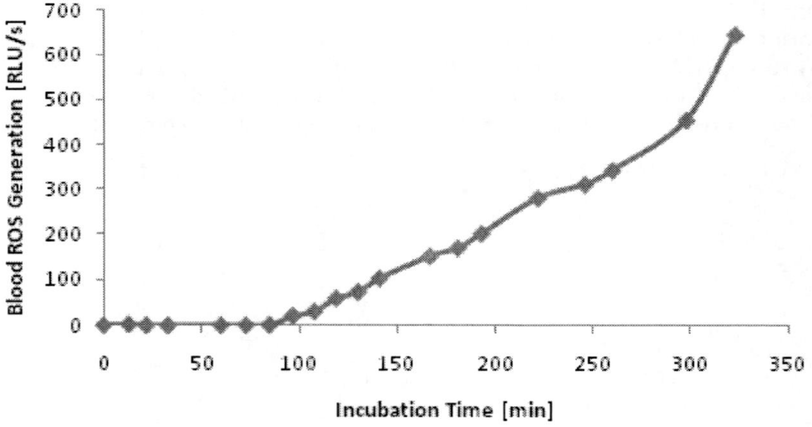

Figure 4. Kinetic of normal ROS generation in citrated blood in BRGA-60-. Normal citrated blood was analyzed in the blood ROS generation assay (BRGA; 125 µl HBSS in black Brand®781608 wells, 10 µl citrated blood, 10 µl 5 mM luminol). At 60 min reaction time 5 µl 18 µg/ml zymosan A were added (BRGA-60-). After 0-323 min (37°C) the light emissions per well were measured by a photons-multiplying microtiter plate photometer (LUmo).

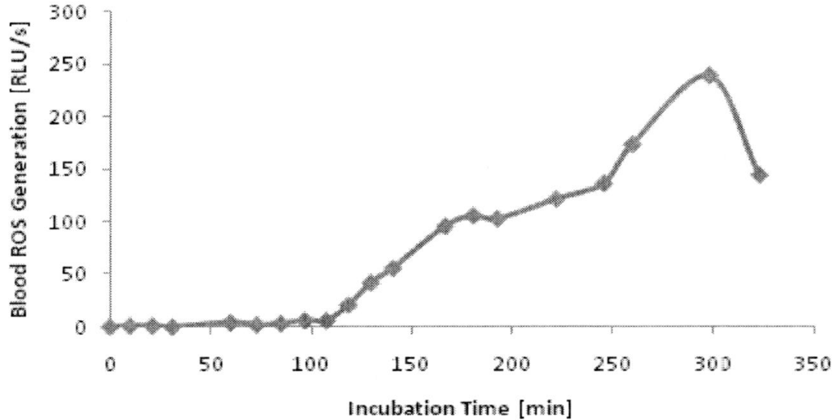

Figure 5. Kinetic of normal ROS generation in EDTA - blood in BRGA-60-. Normal EDTA-blood was analyzed in the blood ROS generation assay (BRGA; 125 µl HBSS in black Brand®781608 wells, 10 µl EDTA-blood, 10 µl 5 mM luminol). At 60 min reaction time 5 µl 18 µg/ml zymosan A were added (BRGA-60-). After 0-323 min (37°C) the light emissions per well were measured by a photons-multiplying microtiter plate photometer (LUmo).

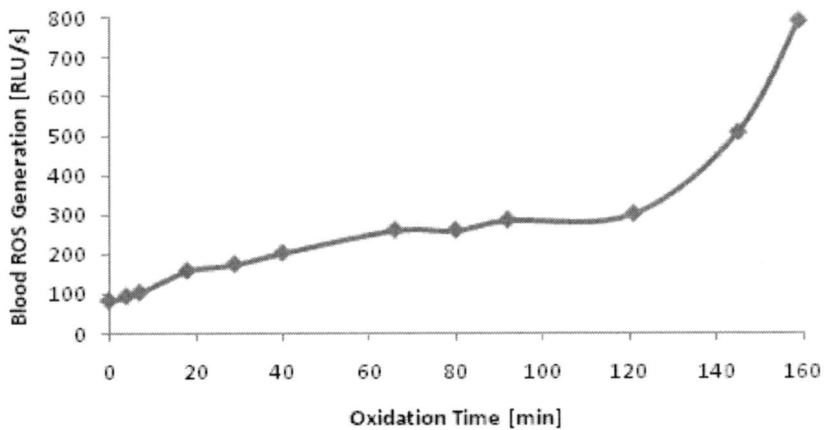

Figure 6. Blood ROS generation in dependence of oxidation time in BRGA-60-. Normal EDTA-blood, oxidized at time point 63 min of figure 5 for 0-159 min (37°C) by 40 µl 0.5 mM chloramine-T®, was analyzed in the BRGA-60- as described in figure 5.

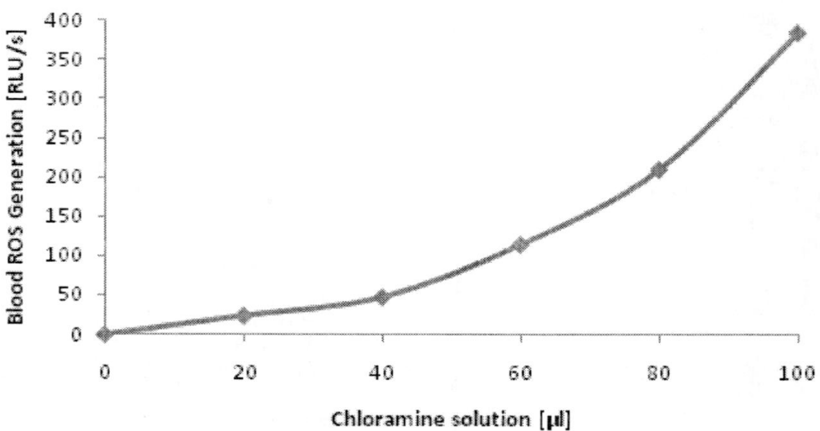

Figure 7. Calibration of LUmo. The analyzer can be calibrated by pure chloramine-T®/luminol. Here 10 µl 5 mM luminol in 0.9 % NaCl were incubated with 0-100 µl 0.5 mM chloramine-T® in 0.9% NaCl, 2.5 mM Na_3-citrate, pH 7.4 in black polystyrene wells. The light emission per well was measured.

Figure 8. Chemical structure of chloramine T® [24]. N-chloro-toluene-sulfonamide (usually in tri-hydrated form of the sodium salt; MW=282 Daltons) contains a delta-negatively charged N-atom. The Cl atom linked to it has some characteristics like chloronium (Cl^+). These positively charged halonium ions could react with negatively charged hydroxyl ions (HO^-) that at neutral pH are present in water in small quantities of about 10^{-7} mol/l = 10^{-4} mmol/l = 0.1 µmol/l. Then $^1O_2^*$ is generated by: $2\,Cl^+ + 2\,HO^- \leftrightarrow {}^1O_2^* + 2\,HCl$.

In conclusion, the present data underline the importance of singlet oxygen /light in cellular fibrinolysis. 6.3 nmoles of CT given to 10 µl citrated blood (equivalent to 0.63 mM) suggests that chloramines in these 0.5-1 mM concentrations in blood are strongly thrombolytic [16]. Here 4-fold this amount of chloramine was added and more than 10-fold enhancement of blood ROS generation appeared, i.e. chloramine concentrations of about 2-4 mM that could well be reached in the micro-environment of activated neutrophils are strongly fibrinolytic, especially in combination with Ca^{2+} chelating inflammatory molecules.

In normal blood about 50% of the NADPH-armamentarium is kept in reserve for continuous activation or following multiple activations, if necessary, e.g. against life-threatening micro-thrombi.

REFERENCES

[1] Stief TW. The physiology and pharmacology of singlet oxygen. *Med Hypoth.* 2003; 60: 567-72.

[2] Stief TW. Regulation of hemostasis by singlet oxygen ($^1\Delta O_2$). *Curr Vasc Pharmacol* 2004; 2: 357-62.

[3] Stief TW, Fareed J. The antithrombotic factor singlet oxygen/light ($^1O_2$/hν). *Clin Appl Thrombosis/Hemostasis* 2000; 6: 22-30.

[4] Stief TW. The blood fibrinolysis / deep - sea analogy: a hypothesis on the cell signals singlet oxygen / photons as natural antithrombotics. *Thromb Res.* 2000; 99: 1-20.

[5] Briviba K, Klotz LO, Sies H. Toxic and signaling effects of photochemically or chemically generated singlet oxygen in biological systems. *Biol Chem.* 1997; 378: 1259-65.

[6] Rosen H, Klebanoff SJ. Formation of singlet oxygen by the myeloperoxidase-mediated antimicrobial system. *J Biol Chem.* 1977; 252: 4803-10.

[7] Weiss SJ, Lampert MB, Test ST. Long-lived oxidants generated by human neutrophils: characterization and bioactivity. *Science.* 1983; 222: 625-8.

[8] Nathan CF. Neutrophil activation on biological surfaces. Massive secretion of hydrogen peroxide in response to products of macrophages and lymphocytes. *J Clin Invest.* 1987; 80: 1550-60.

[9] Kiryu C, Makiuchi M, Miyazaki J, Fujinaga T, Kakinuma K. Physiological production of singlet molecular oxygen in the myeloperoxidase-H_2O_2-chloride system. *FEBS Lett.* 1999; 443: 154-8.

[10] Seguchi H, Kobayashi T. Study of NADPH oxidase-activated sites in human neutrophils. *J Electron Microsc.* 2002; 51: 87-91.

[11] Stief TW. Neutrophil granulocytes in hemostasis. *Hemostasis Laboratory* 2009; 2: 269-89.

[12] Maghzal G, Krause KH, Stocker R, Jaquet V. Detection of reactive oxygen species derived from the family of NOX (NADPH oxidases). *Free Radic Biol Med*. 2012; 53: 1903-18.

[13] Stief T. Light quants of low wave length prime blood neutrophils for ROS generation. *Hemostasis Laboratory* 2013; 6: (issue 4).

[14] Stief T. Blood neutrophils see UV light: 340 nm ultraviolet A stimulates blood ROS generation nearly half as strong as 405 nm violet photons. *Hemostasis Laboratory* 2013; 6 (issue 4).

[15] Stief T. Blood neutrophils alert each other by photons. *Hemostasis Laboratory* 2013; 6 (issue 4).

[16] Stief T, Cimpean CW. Oxidized human albumin stimulates blood ROS generation. *Hemostasis Laboratory* 2013; 6 (issue 4)

[17] Stief T. The routine blood ROS generation assay (BRGA) triggered by typical septic concentrations of zymosan A. *Hemost Lab*. 2013; 6: 89-98.

[18] Stief TW. The fibrinogen antigenic turbidimetric assay (FIATA). The $X^2\bar{x}$ test: the corrected chi-square comparison against the control-mean. *Clin Appl Thrombosis/Hemostasis* 2007; 13: 73-100.

[19] Stief TW. Modulation of granulocyte mediated thrombolysis. *Hemostasis Laboratory* 2008; 1: 77-102.

[20] Stief T. Micro-thrombi stimulate blood ROS generation.*Hemostasis Laboratory* 2013; 6 (issue 2-3)

[21] Goodridge HS, Wolf AJ, Underhill DM. Beta-glucan recognition by the innate immune system. *Immunol Rev*. 2009; 230: 38-50.

[22] Lee JM, Leach ST, Katz T, Day AS, Jaffe A, Ooi CY. Update of faecal markers of inflammation in children with cystic fibrosis. *Mediators Inflamm*. 2012; 2012:948367.

[23] Stief T. ROS generation in diluted whole blood anticoagulated by citrate, EDTA, or heparin. *Hemostasis Laboratory* 2013; 6 (issue 2-3).

[24] www.wikipedia.org.

Chapter 5

SINGLET OXYGEN – OXIDIZED HUMAN ALBUMIN STIMULATES BLOOD ROS GENERATION

Thomas Stief and Codruta Cimpean

ABSTRACT

Background: Albumin is the blood transporter of fatty acids. Unsaturated fatty acids (R_1-C=C-R_2) are important biological targets of excited singlet molecular oxygen ($^1O_2^*$) to generate excited carbonyls (R-C=O^*) that emit light in the spectrum around 400 nm, the exact wave length of the released photons depending on the chemical nature of R and on the blood matrix. These ultraviolet, violet or dark blue photons are seen by blood neutrophils. The neutrophils modulating action of $^1O_2^*$ was an aspect of the present work.

Material and Methods: 40 µl a) 5% human albumin, b) 2.5% human albumin-2.5% human IgG, or c) 0.9% NaCl in black high quality polystyrene microwells (Brand®781608) and 5 µl 106 mM Na_3-citrate, pH 7.4 were oxidized with 0-8.9 mM (final) chloramine-T® (N-chloro-toluene-sulfonamide) for 10 min (37°C). 125 µl Hanks′ Balanced Salt Solution, 10 µl normal citrated blood, and 10 µl 5 mM luminol in 0.9% NaCl were added. The plate was incubated at 37°C in sterile incubator and repetitively measured in a photons-multiplyer microtiter plate luminometer (LUmo). After 33 min (37°C) 20 µl 36 µg/ml zymosan A (ZyA; 2.9 µg/ml final) in 0.9% NaCl were added and the light emissions of the wells were followed in the blood ROS generation assay (BRGA).

Results and Discussion: In BRGA-33-111 maximal blood ROS generations of 9412 RLU/s in unoxidized and 14865 RLU/s in 0.14 mM oxidized human albumin appeared. 7816 RLU/s and 15934 RLU/s were the respective values for the 0.9% NaCl samples. 4128 RLU/s and 3595 RLU/s were the respective values for oxidized human albumin/human IgG, the maximum reached in BRGA-33-85. This means that the singlet oxygen generator chloramine doubles the blood ROS generation in oxidized blood: 6.3 nmoles of CT reacted with 10 µl citrated blood which is equivalent to 0.63 mM final in blood. Singlet oxygen favours cellular fibrinolysis and suggests that chloramines in these 0.5-1 mM concentrations in blood could be exciting new treatment options against lipid enveloped viruses. Up- and down-regulation of NADPH-oxidase seems to be regulated by singlet oxygen and photons in the range of about 380 nm.

Keywords: Reactive oxygen species, ROS, singlet oxygen, chloramine, neutrophils, albumin, unsaturated fatty acids, PUFA, excited carbonyls, UV, violet, dark blue photons, BRGA

INTRODUCTION

Albumin transports fatty acids in the circulating blood [1]. Unsaturated fatty acids (R_1-C=C-R_2), especially polyunsaturated fatty acids (PUFA), are important biological reaction partners of excited singlet molecular oxygen ($^1O_2^*$) yielding excited carbonyls (R-C=O^*) that emit light in the spectrum around 400 nm, the exact wave length of the released photons depend on the chemical nature of R and on the blood matrix [2,3]. Blood neutrophils possess light receptors for these ultraviolet, violet or dark blue photons [4-8]. Phagocytes are the main cells of innate immunology (neutrophils [9], monocytes/macrophages [10], dendritic cells [11]). It is suggested that dark blue photons awaken passive phagocytes, violet photons alert phagocytes, and ultraviolet photons alert them strongly. The neutrophils modulating action of oxidized albumin or of $^1O_2^*$ itself were aspects of the present work. The new blood ROS generation assay (BRGA) is an easy quantitative screening method for any compound that modulates the generation of reactive oxygen species (ROS) by blood neutrophils [12].

Material and Methods

40 µl a) 5% human albumin-0.9% NaCl (Baxter, Unterschleißheim, Germany), b) 2.5% human albumin-2.5% human IgG (Gammagard; Baxter), or c) 0.9% NaCl in black high quality polystyrene flat-bottomed microwells (Brand, Wertheim, Germany; article nr. 781608) and 5 µl 106 mM (12 mM final in oxidation reaction) Na_3-citrate, pH 7.4 were oxidized with 0-8.9 mM (final) chloramine-T® (N-chloro-toluene-sulfonamide; Sigma, Deisenhofen, Germany; MW: 282 Daltons in its usual trihydrate form) for 10 min (37°C) in sterile incubator. 125 µl Hanks′ Balanced Salt Solution (HBSS; modified without phenol red; SAFC Biosciences-Sigma, Deisenhofen, Germany; article nr. 55037C-1000ML), 10 µl normal blood supplemented with 11 mM Na_3-citrate, pH 7.4 (drawn after written informed consent in polypropylene monovettes from Sarstedt, Nümbrecht, Germany; stored for 1d at 23°C), and 10 µl 5 mM luminol sodium salt (Sigma; article nr. A4685-1G; 1g dissolved in 25.1 ml = 200 mM stem solution) in 0.9% NaCl were added. The plate was incubated at 37°C in sterile incubator and repetitively measured in a photons-multiplyer microtiter plate luminometer (LUmo; Autobio labtec – anthos, Krefeld, Germany). After 33 min (37°C) 20 µl 36 µg/ml (2.9 µg/ml final) zymosan A (ZyA; Sigma; article nr. Z-4250-1G, lot nr. 27H0495) in 0.9% NaCl were added and the light emissions of the wells were measured with an integration time of 1s.

HBSS was 185.4 mg/l $CaCl_2 \cdot 2\,H_2O$, 200 mg/l $MgSO_4 \cdot 7\,H_2O$, 400 mg/l KCl, 60 mg/l KH_2PO_4, 350 mg/l $NaHCO_3$, 8000 mg/l NaCl, 90 mg/l Na_2HPO_4, 1000 mg/l glucose, pH 7.0-7.4. Expressed in molarity, the concentrations of the HBSS components are: 1.3 mM Ca^{2+}, 0.8 mM Mg^{2+}, 5.8 mM K^+, 143 mM Na^+, 144 mM Cl^-, 1.6 mM SO_4^{2-}, 0.4 mM $H_2PO_4^-$, 0.6 mM HPO_4^{2-}, 4.2 mM HCO_3^-, 5.6 mM glucose.

Results and Discussion

111 min were needed to obtain maximal blood ROS generations of 9412 RLU/s in unoxidized human albumin and 14865 RLU/s in 0.14 mM oxidized human albumin (Figure 1a). 7816 RLU/s and 15934 RLU/s were the respective values for the 0.9% NaCl sample (Figure 1b). 4128 RLU/s and 3595 RLU/s were the respective values for oxidized albumin - IgG, the maximum reached after 85 min (Figure 1c). This means that the singlet oxygen generator chloramine doubles the blood ROS generation in oxidized

blood: 6.3 nmoles of CT reacted with 10 µl citrated blood (equivalent to 0.63 mM) since CT in physiol. NaCl is very stable [13]. This gives further evidence that singlet oxygen favours thrombolysis [14-16] and suggests that chloramines in these 0.5-1 mM concentrations in blood [17,18] could be exciting new treatment options against the AIDS virus or other enveloped viruses [19-21] or against severe thrombosis.

Up to 1.1 mM chloramine oxidation of 5% albumin the BRGA-33-62 activity increased slightly (Figure 2). 3 mM chloramine is the approx. IC50 on BRGA-33-62. When compared with 5% albumin 2.5% albumin+2.5% IgG suppressed about 50% of the unoxidized control BRGA. Up to 2.2 mM chloramine oxidation of 2.5% albumin+2.5% IgG there was no major change of blood ROS generation; the approx. IC50 was 3.3 mM chloramine. Only 0.1 mM chloramine added to 0.9 % NaCl resulted in an approx. SC150, 1 mM chloramine was the approx. IC50 on BRGA-33-62 (Figure 2).

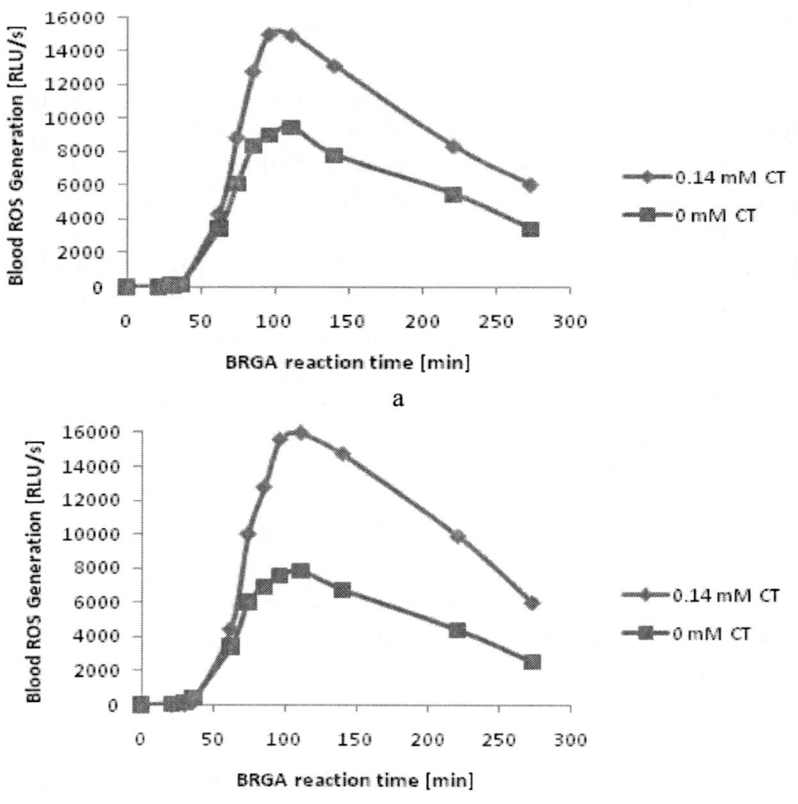

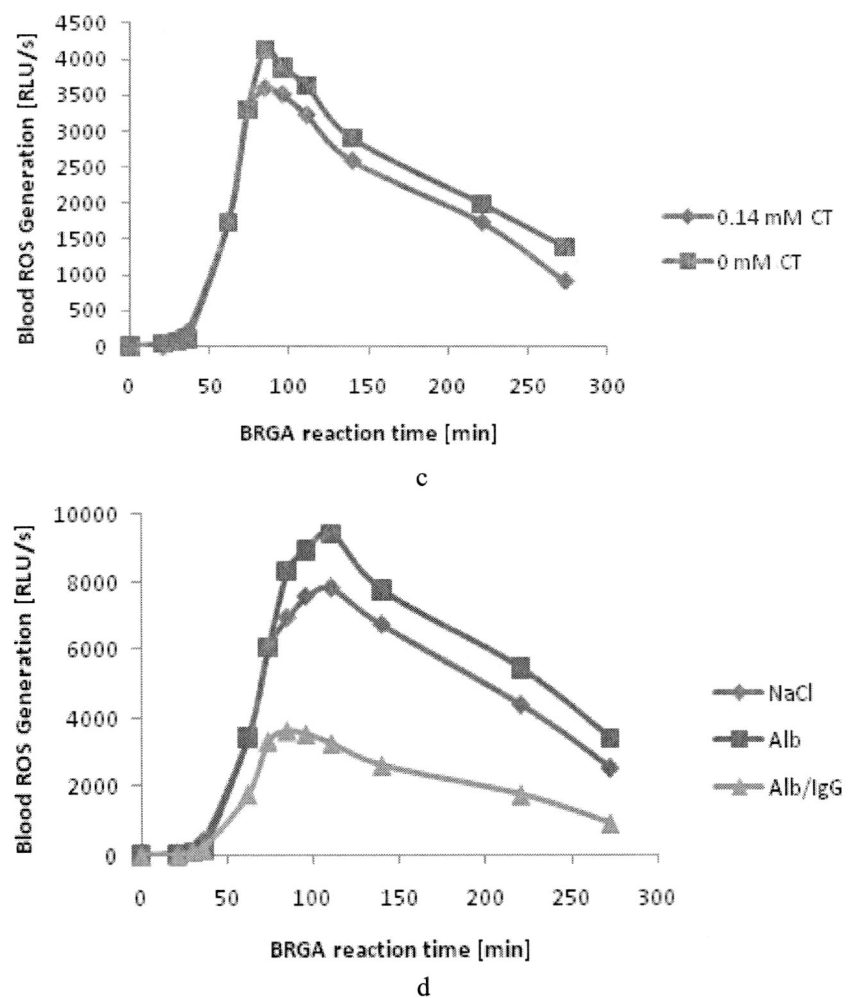

Figure 1. Blood ROS generation kinetic. 40 µl 5% human albumin (Figure 1a), 0.9% NaCl (Figure 1b), or 2.5% human albumin-2.5% human IgG (Figure 1c) and 5 µl 106 mM Na_3-citrate, pH 7.4, in black microwells (Brand®781608) were oxidized with 0.14 mM chloramine T® (CT; controls: 0 mM CT) for 10 min (37°C). 125 µl Hanks′ Balanced Salt Solution (HBSS), 10 µl normal citrated blood, and 10 µl 5 mM luminol in 0.9% NaCl were added and pre-reacted for 33 min (37°C). 2.9 µg/ml zymosan A (ZyA) in 0.9% NaCl were added and the light emissions of the wells in relative light units per second (RLU/s) were measured in the blood ROS generation assay (BRGA) with reaction times up to 273 min (BRGA-33-273). Figure 1d: comparison of the 3 samples without addition of the singlet oxygen generator CT. Intra-assay CVs < 5%.

Figures 3-9 demonstrate further BRGA data with the same pre-reaction time of 33 min and different reaction times of 74 to 273 min (BRGA-33-74 to BRGA-33-273). The approx. SC150 of chloramine on albumin-oxidation induced stimulation of blood ROS generation at BRGA times ≥ 85 min was only about 0.03-0.06 mM (Figure 10). At 85-111 min reaction time (immediately before or at the ROS maximum) there appeared even approx. SC200 values of 0.6 mM, 0.3 mM, or 0.3 mM for BRGA-33-85, BRGA-33-96, or BRGA-33-111, respectively.

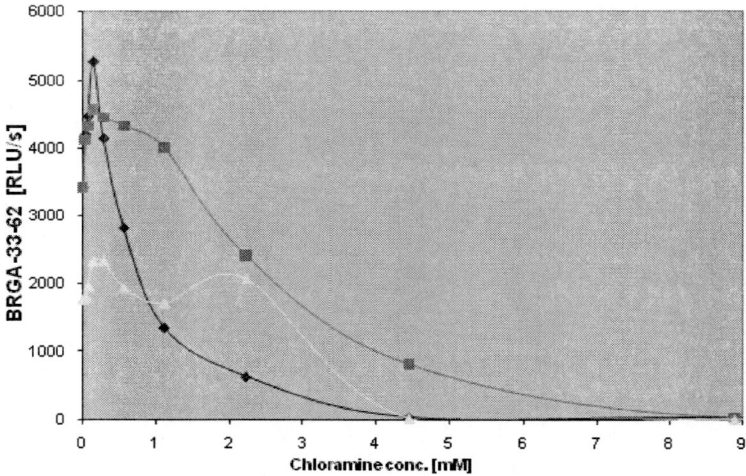

Figure 2. Blood ROS generation modulation by singlet oxygen in BRGA-33-62. 40 µl 5% human albumin (■), 2.5% human albumin-2.5% human IgG (Δ), or 0.9% NaCl (♦) in black microwells (Brand®781608) and 5 µl 106 mM Na_3-citrate, pH 7.4 were oxidized with 0-8.9 mM (final) chloramine-T® for 10 min (37°C). 125 µl Hanks′ Balanced Salt Solution, 10 µl normal citrated blood, and 10 µl 5 mM luminol in 0.9% NaCl were added. After 33 min (37°C) preincubation 2.9 µg/ml zymosan A (ZyA) in 0.9% NaCl were added and the light emissions of the wells in relative light units per second (RLU/s) were measured in the blood ROS generation assay (BRGA), here at 62 min main reaction time (BRGA-33-62).

For the action of singlet oxygen on citrated blood approx. SC150 values of less than 0.04 mM chloramine in NaCl resulted, the longer the BRGA reaction time the lower the approx. SC150 (Figure 11). It seems that especially at prolonged neutrophil stress the cells communicate by UV-photons and become partially resistant to further down-regulation of cellular thrombolysis [6-8,16].

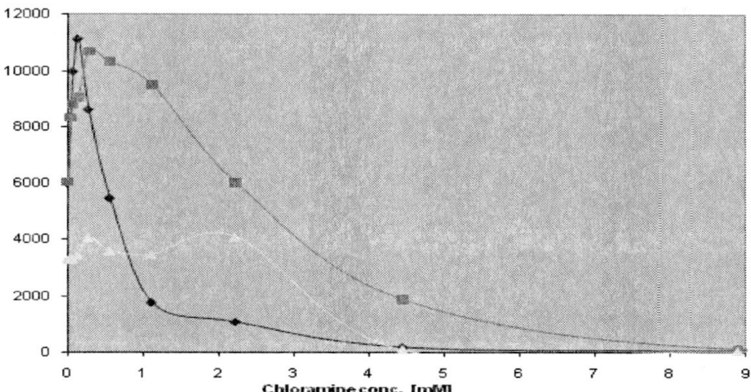

Figure 3. Blood ROS generation modulation by singlet oxygen in BRGA-33-74. 40 µl 5% human albumin (■), 2.5% human albumin-2.5% human IgG (Δ), or 0.9% NaCl (♦) in black microwells (Brand®781608) and 5 µl 106 mM Na_3-citrate, pH 7.4 were oxidized with 0-8.9 mM (final) chloramine-T® for 10 min (37°C). 125 µl HBSS Solution, 10 µl normal citrated blood, and 10 µl 5 mM luminol were added. After 33 min (37°C) preincubation 2.9 µg/ml zymosan A (ZyA) in 0.9% NaCl were added and the RLU/s were measured in the blood ROS generation assay (BRGA), here at 74 min main reaction time (BRGA-33-74).

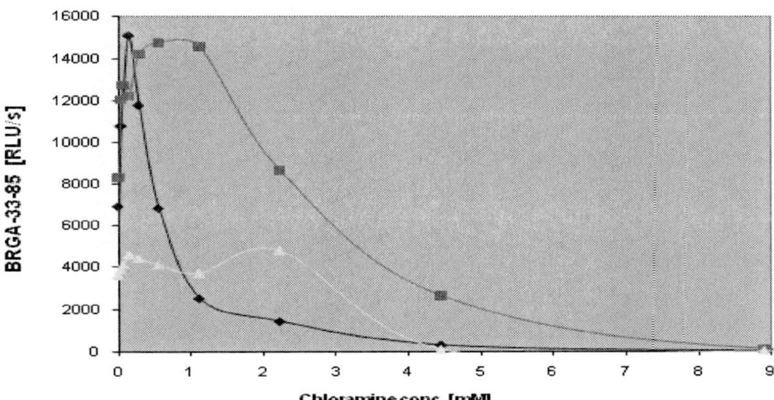

Figure 4. Blood ROS generation modulation by singlet oxygen in BRGA-33-85. 40 µl 5% human albumin (■), 2.5% human albumin-2.5% human IgG (Δ), or 0.9% NaCl (♦) in black microwells (Brand®781608) and 5 µl 106 mM Na_3-citrate, pH 7.4 were oxidized with 0-8.9 mM (final) chloramine-T® for 10 min (37°C). 125 µl HBSS, 10 µl normal citrated blood, and 10 µl 5 mM luminol in 0.9% NaCl were added. After 33 min (37°C) preincubation 2.9 µg/ml ZyA in 0.9% NaCl were added and the RLU/s were measured in the blood ROS generation assay (BRGA), here at 85 min main reaction time (BRGA-33-85).

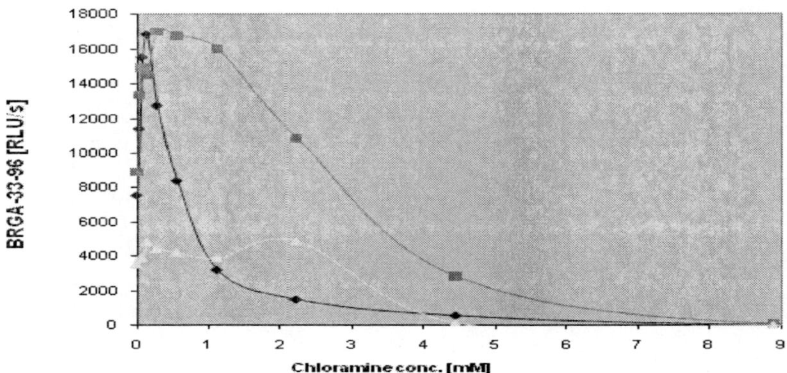

Figure 5. Blood ROS generation modulation by singlet oxygen in BRGA-33-96. 40 μl 5% human albumin (■), 2.5% human albumin-2.5% human IgG (Δ), or 0.9% NaCl (♦) in black microwells (Brand®781608) and 5 μl 106 mM Na_3-citrate, pH 7.4 were oxidized with 0-8.9 mM (final) chloramine-T® for 10 min (37°C). 125 μl HBSS, 10 μl normal citrated blood, and 10 μl 5 mM luminol in 0.9% NaCl were added. After 33 min (37°C) preincubation 2.9 μg/ml ZyA in 0.9% NaCl were added and the RLU/s were measured in the blood ROS generation assay (BRGA), here at 96 min main reaction time (BRGA-33-96).

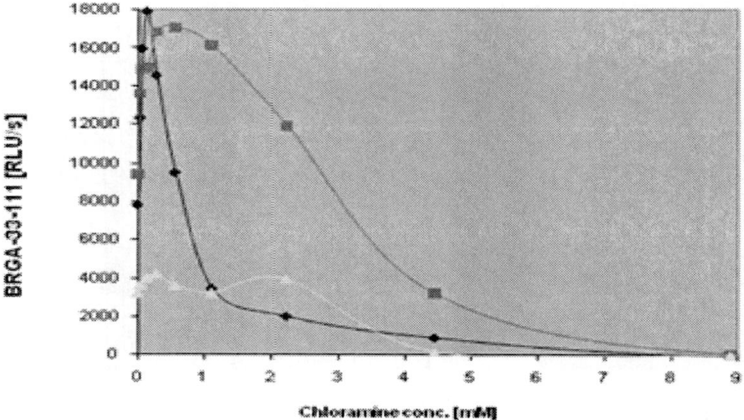

Figure 6. Blood ROS generation modulation by singlet oxygen in BRGA-33-111. 40 μl 5% human albumin (■), 2.5% human albumin-2.5% human IgG (Δ), or 0.9% NaCl (♦) in black microwells (Brand®781608) and 5 μl 106 mM Na_3-citrate, pH 7.4 were oxidized with 0-8.9 mM (final) chloramine-T® for 10 min (37°C). 125 μl HBSS, 10 μl normal citrated blood, and 10 μl 5 mM luminol in 0.9% NaCl were added. After 33 min (37°C) preincubation 2.9 μg/ml ZyA in 0.9% NaCl were added and the RLU/s were measured in the blood ROS generation assay (BRGA), here at 111 min main reaction time (BRGA-33-111).

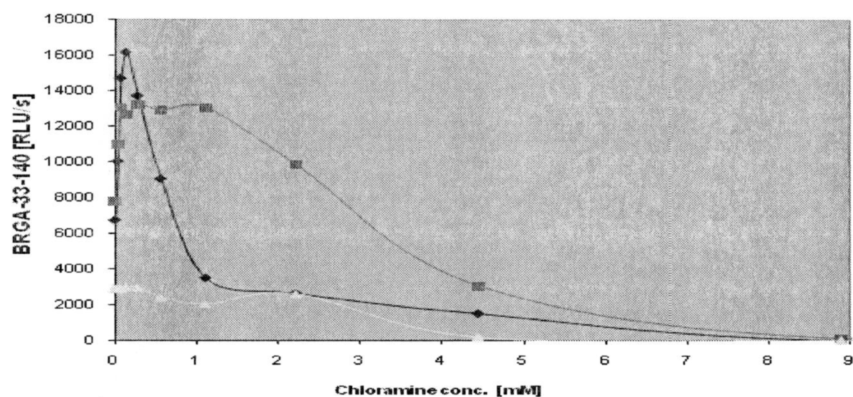

Figure 7. Blood ROS generation modulation by singlet oxygen in BRGA-33-140. 40 μl 5% human albumin (■), 2.5% human albumin-2.5% human IgG (Δ), or 0.9% NaCl (♦) in black microwells (Brand®781608) and 5 μl 106 mM Na_3-citrate, pH 7.4 were oxidized with 0-8.9 mM (final) chloramine-T® for 10 min (37°C). 125 μl HBSS, 10 μl normal citrated blood, and 10 μl 5 mM luminol in 0.9% NaCl were added. After 33 min (37°C) preincubation 2.9 μg/ml ZyA in 0.9% NaCl were added and the RLU/s were measured in the blood ROS generation assay (BRGA), here at 140 min main reaction time (BRGA-33-140).

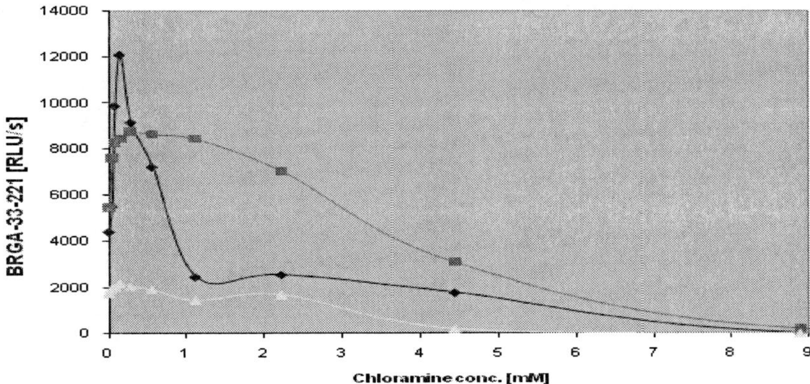

Figure 8. Blood ROS generation modulation by singlet oxygen in BRGA-33-221. 40 μl 5% human albumin (■), 2.5% human albumin-2.5% human IgG (Δ), or 0.9% NaCl (♦) in black microwells (Brand®781608) and 5 μl 106 mM Na_3-citrate, pH 7.4 were oxidized with 0-8.9 mM (final) chloramine-T® for 10 min (37°C). 125 μl HBSS, 10 μl normal citrated blood, and 10 μl 5 mM luminol in 0.9% NaCl were added. After 33 min (37°C) preincubation 2.9 μg/ml ZyA in 0.9% NaCl were added and the RLU/s were measured in the blood ROS generation assay (BRGA), here at 221 min main reaction time (BRGA-33-221).

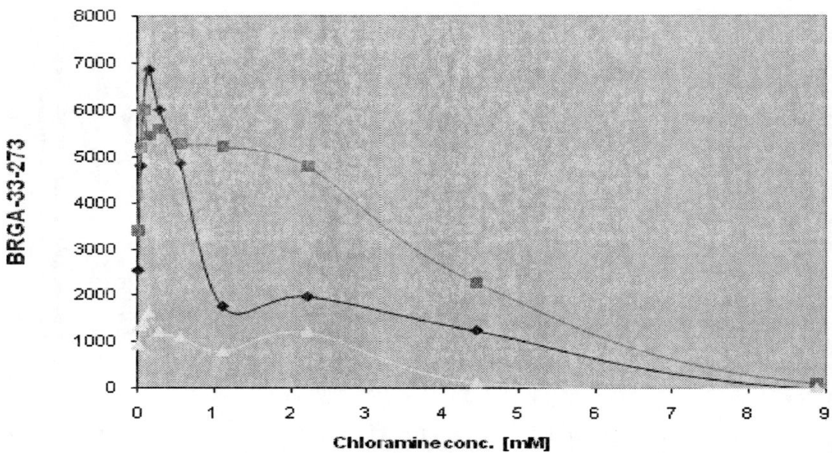

Figure 9. Blood ROS generation modulation by singlet oxygen in BRGA-33-273. 40 µl 5% human albumin (■), 2.5% human albumin-2.5% human IgG (Δ), or 0.9% NaCl (♦) in black microwells (Brand®781608) and 5 µl 106 mM Na_3-citrate, pH 7.4 were oxidized with 0-8.9 mM (final) chloramine-T® for 10 min (37°C). 125 µl HBSS, 10 µl normal citrated blood, and 10 µl 5 mM luminol in 0.9% NaCl were added. After 33 min (37°C) preincubation 2.9 µg/ml ZyA in 0.9% NaCl were added and the RLU/s were measured in the blood ROS generation assay (BRGA), here at 273 min main reaction time (BRGA-33-273).

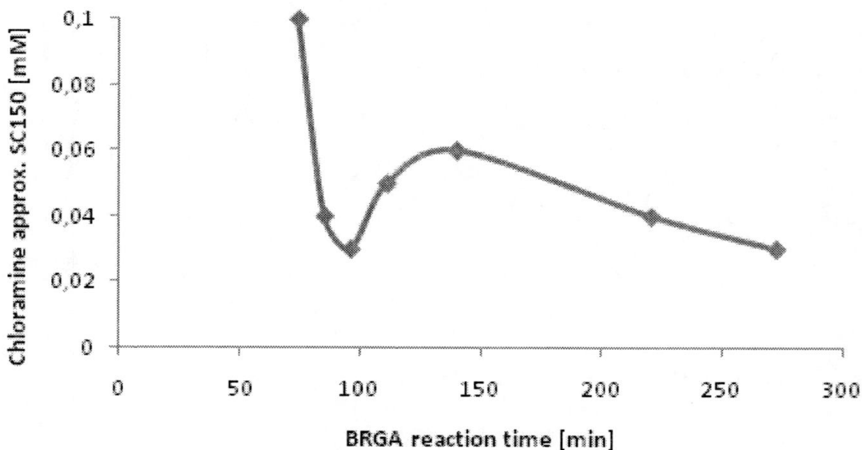

Figure 10. SC150 values of chloramine on albumin oxidation. Albumin was oxidized by the singlet oxygen generator chloramine T®. The oxidized albumin samples were analyzed in the BRGA as described in figure 1. The approximate 150% stimulatory chloramine concentrations (approx. SC150) on blood ROS generation were determined in figures 2-9.

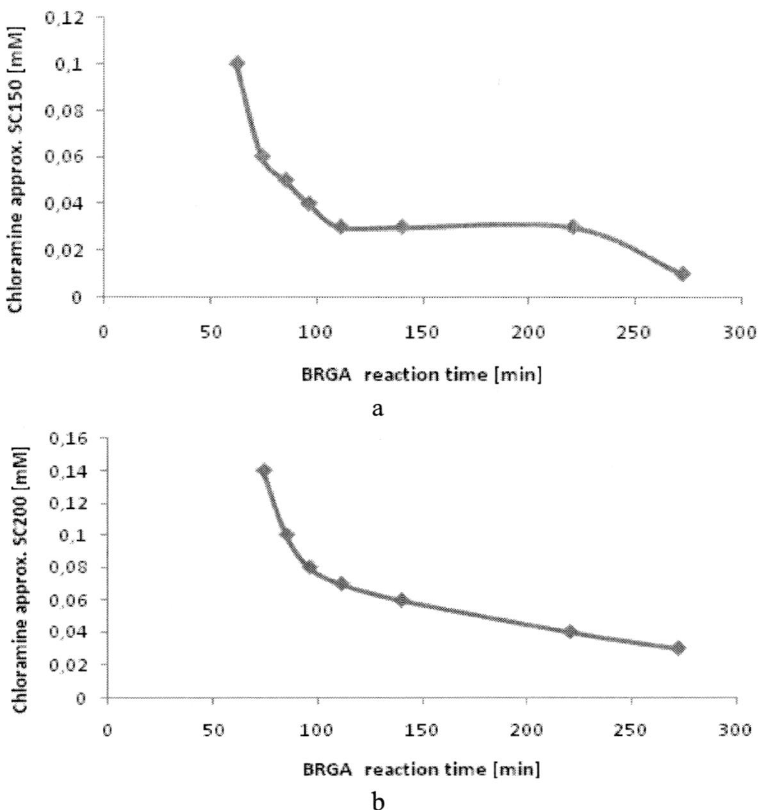

Figure 11. Stimulation of blood ROS generation by singlet oxygen. The approximate 150% (Figure 11a) or 200% stimulatory chloramine concentrations (Figure 11b) on blood ROS generation were determined in figures 2-9. Under the reaction conditions used chloramine T® exclusively generates the non-radicalic excited singlet oxygen [17,18,45].

The longer the blood ROS generation lasts the more chloramine is needed to suppress it (high dose inactivating action of chloramines), i.e. the higher the resulting approx. IC50 of chloramine, there appeared a 4-fold increase with the NaCl controls (from 1 to 4 mM) and a 2-fold increase with albumin (from 3 to 6 mM). The albumin/IgG mixture′s IC50 was about 3.5 mM and resistant to BRGA reaction time (Figure 12). It is suggested that singlet oxygen switches on the down-regulation of NADPH-oxidase and that this switch is gradually lost with increasing blood ROS stress.

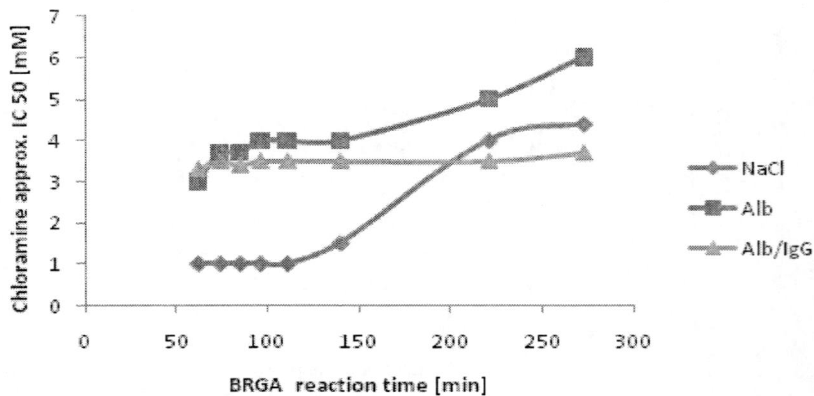

Figure 12. IC50 values of chloramine on sample oxidation. 5% Albumin, 2.5% albumin-2.5% IgG, or 0.9% NaCl (control) were incubated for 10 min (37°C) with chloramine T®. The samples were then analyzed in the BRGA as described in figure 1. The approx. IC50 of chloramine on blood ROS generation were determined in figures 2-9.

Figure 13. Chemical structure of chloramine T® [44]. N-chloro-toluene-sulfonamide (usually in tri-hydrated form of the sodium salt; MW=282 Daltons) contains a delta-negatively charged N-atom. The Cl atom linked to it has some characteristics like positively charged Cl ions, so-called chloronium ions (Cl^+). These positively charged halonium ions could react with HO^- that at neutral pH are always present in water (in small quantities of about 10^{-7} mol/l $=10^{-4}$ mmol/l = 0.1 µmol/l). Then $^1O_2$ is generated: $2\ Cl^+ + 2\ HO^- \leftrightarrow {}^1O_2 + 2\ HCl$. According to the law of mass action of Guldberg and Waage $k = [a]*[b] / [c]*[d]$ chloramines are very stable as long as the reactive end product of the reaction $^1O_2$ is not consumed by oxidizable compounds in the matrix (entire environment of $^1O_2$ including fluid and solid phase), i.e. as long as the concentration of Cl^+ [a] (stabilized by 3 H_2O and bound to the amine) and that of $^1O_2$ [c] is constant.

In conclusion, oxidized albumin stimulates blood ROS generation. This could be a direct action of singlet oxygen on the protein [22,23] but oxidation of the fatty acid transporter albumin might well attack unsaturated fatty acids that are transformed in excited carbonyls [24-28]. The physiology and pathophysiology of inflammation is complex [29,30] with many up-regulating [31-38] and several down-regulating compounds [39,40]. There is increasing evidence that inflammation is linked to photons in the wavelength range around 380 nm [41,42]. One major future focus will be the physiology of chloramine [13,43] that in physiological conditions only generates the non-radicalic singlet oxygen [44,45].

REFERENCES

[1] Eberhart LH, Graf J, Morin AM, Stief T, Kalder M, Lattermann R, Schricker T. Randomised controlled trial of the effect of oral premedication with dexamethasone on hyperglycaemic response to abdominal hysterectomy. *Eur J Anaesthesiol.* 2011; 28: 195-201.

[2] Stief TW. The physiology and pharmacology of singlet oxygen. *Med Hypoth.* 2003; 60: 567-72.

[3] Stief TW. Regulation of hemostasis by singlet-oxygen ($^1\Delta O_2$). *Curr Vasc Pharmacol* 2004; 2: 357-62.

[4] Stief TW, Fareed J. The antithrombotic factor singlet oxygen/light ($^1O_2$/hv). *Clin Appl Thrombosis/Hemostasis* 2000; 6: 22-30.

[5] Stief TW. The blood fibrinolysis / deep - sea analogy: a hypothesis on the cell signals singlet oxygen / photons as natural antithrombotics. *Thromb Res.* 2000; 99: 1-20.

[6] Stief T. Light quants of low wave length prime blood neutrophils for ROS generation. *Hemostasis Laboratory* 2013; 6: (issue 4).

[7] Stief T. Blood neutrophils see UV light: 340 nm ultraviolet A stimulates blood ROS generation nearly half as strong as 405 nm violet photons. *Hemostasis Laboratory* 2013; 6 (issue 4).

[8] Stief T. Blood neutrophils alert each other by photons. *Hemostasis Laboratory* 2013; 6 (issue 4).

[9] Stief TW. Neutrophil granulocytes in hemostasis.*Hemostasis Laboratory* 2009; 2: 269-89.

[10] Ano Y, Sakudo A, Onodera T. Role of microglia in oxidative toxicity associated with encephalomyocarditis virus infection in the central nervous system. *Int J Mol Sci.* 2012; 13: 7365-74.

[11] Savina A, Vargas P, Guermonprez P, Lennon AM, Amigorena S. Measuring pH, ROS production, maturation, and degradation in dendritic cell phagosomes using cytofluorometry-based assays. *Methods Mol Biol.* 2010; 595: 383-402.

[12] Stief T. The routine blood ROS generation assay (BRGA) triggered by typical septic concentrations of zymosan A. Hemost Lab. 2013; 6: 89-98.

[13] Weiss SJ, Lampert MB, Test ST. Long-lived oxidants generated by human neutrophils: characterization and bioactivity. *Science*. 1983; 222: 625-8.

[14] Stief TW. Singlet oxygen potentiates thrombolysis. *Clin Appl Thrombosis/Hemostasis* 2007; 13: 259-78.

[15] Stief TW. Modulation of granulocyte mediated thrombolysis. *Hemostasis Laboratory* 2008; 1: 77-102.

[16] Stief T. Micro-thrombi stimulate blood ROS generation. *Hemostasis Laboratory* 2013; 6 (issue 2-3)

[17] Stief TW. Singlet oxygen and thrombin generation: 0.5-1 mM chloramine as anti-viral therapy. *Hemostasis Laboratory* 2010; 3: 311-24.

[18] Stief TW. Singlet oxygen and thrombin generation in RECA. *Hemostasis Laboratory* 2011; 4: 237-54.

[19] Stief TW. Singlet oxygen – oxidizable lipids in the HIV membrane, new targets for AIDS therapy ? *Med Hypoth.* 2003; 60: 575-7.

[20] Stief TW. Hemostasis tolerable singlet oxygen - a perspective in AIDS therapy. *Hemostasis Laboratory* 2008; 1: 21-40.

[21] Stief T. AIDS drugs might work via stimulation of singlet oxygen generating neutrophils. *Hemostasis Laboratory* 2013; 6 (issue 2-3)

[22] Summers FA, Forsman Quigley A, Hawkins CL. Identification of proteins susceptible to thiol oxidation in endothelial cells exposed to hypochlorous acid and N-chloramines. *Biochem Biophys Res Commun.* 2012; 425: 157-61.

[23] Candiano G, Petretto A, Bruschi M, Santucci L, Dimuccio V, Prunotto M, Gusmano R, Urbani A, Ghiggeri GM. The oxido-redox potential of albumin methodological approach and relevance to human diseases. *J Proteomics*. 2009; 73: 188-95.

[24] Ralay Ranaivo H, Hodge JN, Choi N, Wainwright MS. Albumin induces upregulation of matrix metalloproteinase-9 in astrocytes via MAPK and reactive oxygen species-dependent pathways. *J Neuroinflammation*. 2012; 9: 68.

[25] Ralay Ranaivo H, Wainwright MS. Albumin activates astrocytes and microglia through mitogen-activated protein kinase pathways. *Brain Res.* 2010; 1313: 222-31.

[26] Lee JY, Chang JW, Yang WS, Kim SB, Park SK, Park JS, Lee SK. Albumin-induced epithelial-mesenchymal transition and ER stress are regulated through a common ROS-c-Src kinase-mTOR pathway: effect of imatinib mesylate. *Am J Physiol Renal Physiol.* 2011; 300: F1214-22.

[27] Kanno S, Kurauchi K, Tomizawa A, Yomogida S, Ishikawa M. Albumin modulates docosahexaenoic acid-induced cytotoxicity in human hepatocellular carcinoma cell lines. *Toxicol Lett.* 2011; 200: 154-61.

[28] Nøding R, Schønberg SA, Krokan HE, Bjerve KS. Effects of polyunsaturated fatty acids and their n-6 hydroperoxides on growth of five malignant cell lines and the significance of culture media. *Lipids.* 1998; 33: 285-93.

[29] Ježek J, Jabůrek M, Zelenka J, Ježek P. Mitochondrial phospholipase A_2 activated by reactive oxygen species in heart mitochondria induces mild uncoupling. *Physiol Res.* 2010; 59: 737-47.

[30] Huang X, Zhang J, Liu J, Sun L, Zhao H, Lu Y, Wang J, Li J. C-reactive protein promotes adhesion of monocytes to endothelial cells via NADPH oxidase-mediated oxidative stress. *J Cell Biochem.* 2012; 113: 857-67.

[31] Goodridge HS, Wolf AJ, Underhill DM. Beta-glucan recognition by the innate immune system. *Immunol Rev.* 2009; 230: 38-50.

[32] Nathan CF. Neutrophil activation on biological surfaces. Massive secretion of hydrogen peroxide in response to products of macrophages and lymphocytes. *J Clin Invest.* 1987; 80: 1550-60.

[33] Seguchi H, Kobayashi T. Study of NADPH oxidase-activated sites in human neutrophils. *J Electron Microsc.* 2002; 51: 87-91.

[34] Rosen H, Klebanoff SJ. Formation of singlet oxygen by the myeloperoxidase-mediated antimicrobial system. *J Biol Chem.* 1977; 252: 4803-10.

[35] Kiryu C, Makiuchi M, Miyazaki J, Fujinaga T, Kakinuma K. Physiological production of singlet molecular oxygen in the myeloperoxidase-H_2O_2-chloride system. *FEBS Lett.* 1999; 443: 154-8.

[36] Maghzal G, Krause KH, Stocker R, Jaquet V. Detection of reactive oxygen species derived from the family of NOX (NADPH oxidases). *Free Radic Biol Med.* 2012; 53: 1903-18.

[37] Briviba K, Klotz LO, Sies H. Toxic and signaling effects of photochemically or chemically generated singlet oxygen in biological systems. *Biol Chem.* 1997; 378: 1259-65.

[38] Joos C, Marrama L, Polson HE, Corre S, Diatta AM, Diouf B, Trape JF, Tall A, Longacre S, Perraut R. Clinical protection from falciparum malaria correlates with neutrophil respiratory bursts induced by merozoites opsonized with human serum antibodies. *PLoS One*. 2010; 5: e9871.

[39] Klaesson S, Ringdén O, Markling L, Remberger M, Lundkvist I. Immune modulatory effects of immunoglobulins on cell-mediated immune responses in vitro. *Scand J Immunol.* 1993; 38: 477-84.

[40] Semple JW, Kim M, Hou J, McVey M, Lee YJ, Tabuchi A, Kuebler WM, Chai ZW, Lazarus AH.. Intravenous immunoglobulin prevents murine antibody-mediated acute lung injury at the level of neutrophil reactive oxygen species (ROS) production. *PLoS One.* 2012; 7: e31357.

[41] Lavi R, Shainberg A, Shneyvays V, Hochauser E, Isaac A, Zinman T, Friedmann H, Lubart R. Detailed analysis of reactive oxygen species induced by visible light in various cell types. *Lasers Surg Med.* 2010; 42: 473-80.

[42] Lavi R, Ankri R, Sinyakov M, Eichler M, Friedmann H, Shainberg A, Breitbart H, Lubart R. The plasma membrane is involved in the visible light-tissue interaction. *Photomed Laser Surg.* 2012; 30:14-9.

[43] Gottardi W, Nagl M. N-chlorotaurine, a natural antiseptic with outstanding tolerability. *J Antimicrob Chemother.* 2010; 65: 399-409.

[44] www.wikipedia.org

[45] Stief TW. The fibrinogen antigenic turbidimetric assay (FIATA). The $X^2\bar{x}$ test: the corrected chi-square comparison against the control-mean. *Clin Appl Thrombosis/Hemostasis* 2007; 13: 73-100.

Chapter 6

HUMAN IGG CAN MODULATE BLOOD ROS GENERATION

ABSTRACT

Background: Theapeutic immunoglobulins of the G-class (IgG) are given in clinical situations of decreased IgG-concentrations in blood or to suppress hyper-activated phagocytes in autoimmunity. The new blood ROS generation assay (BRGA) is an easy screening method for all types of drugs that might modulate reactive oxygen species (ROS) generation by blood neutrophils. Here 5 different *i.v.* therapeutic immunoglobulins were analyzed in the BRGA.

Material and Methods: 40 µl 0-10.3 g/l (final) privigen® (Behring), Gammagard® (Baxter), Intratect® (Biotest), Kiovig® (Baxter), Gammunex® (Talecris) in 0.9% NaCl (all IgG drugs in -30°C frozen/ 23°C thawed aliquots of 50 g/l) in black Brand®781608 high quality polystyrene microwells were incubated in triplicate with 125 µl Hanks′ balanced salt solution (HBSS; modified without phenol red), 10 µl 0.26 mM luminol (final), 10 µl 1.9 µg/ml (final) zymosan A in 0.9 % NaCl, and 10 µl normal citrated blood or EDTA-blood. The plates were measured within 0-110 min (37°C) in a photons-multiplyer microtiter plate luminometer (LUmo). Approximate 50% inhibitory concentrations (approx. IC50) or approximate 200% stimulatory concentrations (approx. SC200) of IgG on blood ROS generation were determined if present.

Results: Privigen® and Gammagard® were respective blood ROS generation the best IgG drugs. At 25 min incubation time they did not change blood ROS generation. By contrast, Gammunex® and Kiovig® enhanced blood ROS generation two-fold (approx. SC200) at only 0.3 g/l added IgG. Intratect® had an approx. SC200 of 3 g/l added IgG.

Discussion: An ideal IgG drug should a) function as IgG, b) not stimulate blood ROS generation, c) not inhibit blood ROS generation too strongly, d) not act as an activator of AM-coagulation (coagulation triggered by altered matrix). From all the data on points b) to d) together it seems that privigen® could be the best *i.v.* human IgG drug currently on the medical market. Standardized large studies on blood ROS generation modulation by different IgG drugs are required. Exciting times lie ahead of all medical researchers that focus on the biochemistry, pathobiochemistry, physiology, pathophysiology, pharmacology, toxicology, histology, histopathology of singlet oxygen.

Keywords: Reactive oxygen species, ROS, singlet oxygen, neutrophils, therapeutic immunoglobulins, IgG, IVIG

INTRODUCTION

Intravenous immunoglobulins of the G-class (IVIG) in the conc. range of about 5 g/l are indicated in clinical situations of decreased plasmatic IgG-concentrations or to suppress autoimmunity [1-5], for example the phagocyte´s Fc receptor suppressing IVIG can be a treatment option in autoimmune hemolytic anemia [6]. Phagocytes are the main cells of innate immunology (neutrophils [7], monocytes/macrophages [8], dendritic cells [9]). The new blood ROS generation assay (BRGA) is an easy screening method for any drug that might modulate the generation of reactive oxygen species (ROS) by blood neutrophils [10,11]. Here the action 5 different *i.v.* therapeutic immunoglobulins on blood ROS generation was quantified.

MATERIAL AND METHODS

40 µl 0-10.3 g/l (final) privigen® (CSL Behring, Marburg, Germany), Gammagard® (Baxter, Unterschleißheim, Germany), Intratect® (Biotest, Dreieich, Germany), Kiovig® (Baxter, Vienna, Austria), Gammunex® (Talecris, Frankfurt, Germany) in 0.9% NaCl (all IgG drugs in -30°C frozen/ 23°C thawed aliquots of 50 g/l) in black high quality polystyrene microwells (Brand, Wertheim, Germany; article nr. 781608) were incubated in triplicate with 125 µl Hanks´ balanced salt solution (HBSS; modified without phenol red; SAFC Biosciences-Sigma, Deisenhofen, Germany; article nr. 55037C-1000ML), 10 µl samples of normal venous blood anticoagulated by 11 mM

Na_3-citrate or other normal blood samples anticoagulated by 1.6 mg/ml K_3-EDTA (drawn after written informed consent into polypropylene monovettes from Sarstedt, Nümbrecht, Germany and stored for 1d at 23°C prior to analysis), 10 µl 5 mM luminol sodium salt (0.26 mM final) in 0.9 % NaCl (Sigma; article nr. A4685-1G; 1g dissolved in 25.1 ml = 200 mM stem solution), and 10 µl blood ROS generation trigger 1.9 µg/ml (final) zymosan A (ZyA; Sigma; article nr. Z-4250-1G, lot nr. 27H0495). The plates were measured within 0-110 min (37°C) in a photons-multiplyer microtiter plate luminometer (LUmo). Approximate 50% inhibitory concentrations (approx. IC50) or approximate 200% stimulatory conc.(approx. SC200) of IgG on blood ROS generation were determined.

HBSS was 185.4 mg/l $CaCl_2 \cdot 2 H_2O$, 200 mg/l $MgSO_4 \cdot 7 H_2O$, 400 mg/l KCl, 60 mg/l KH_2PO_4, 350 mg/l $NaHCO_3$, 8000 mg/l NaCl, 90 mg/l Na_2HPO_4, 1000 mg/l glucose, pH 7.0-7.4. Expressed in molarity, the concentrations of the HBSS components are: 1.3 mM Ca^{2+}, 0.8 mM Mg^{2+},5.8 mM K^+, 143 mM Na^+, 144 mM Cl^-, 1.6 mM SO_4^{2-}, 0.4 mM $H_2PO_4^-$, 0.6 mM HPO_4^{2-} , 4.2 mM HCO_3^- , 5.6 mM glucose.

RESULTS

The maximal blood ROS generation in citrated blood was about 3000 RLU/s, reached at about 90 min (37°C). Halfmaximal values were obtained after about 60 min (37°C) (Figure 1).

The maximal blood ROS generation in EDTA blood was about 2000 RLU/s, reached at about 50-60 min (37°C). Halfmaximal values were obtained after about 40 min (37°C) (Figure 2).

Privigen® and Gammagard® behaved respective blood ROS generation as the best IgG drugs. At 25 min incubation time they did not change blood ROS generation (Figure 3). By contrast, Gammunex® and Kiovig® enhanced blood ROS generation two-fold (approx. SC200) at only 0.3 g/l added IgG. Intratect® had an approx. SC200 of 3 g/l added IgG.

In EDTA blood analyzed at 25 min privigen®, Gammagard®, and Intratect® had no approx. SC200 value. Gammunex® had an approx. SC200 of 0.3 g/l and Kiovig® one of 0.6 g/l added IgG (Figure 4).

At 38 min reaction time Gammagard®, Intratect®, and Kiovig® had no approx. SC200 or approx. IC50, privigen® had an approx. IC50 of 4 g/l added IgG, Gammunex® had an approx. SC200 of 4 g/l added IgG (Figure 5).

At 59 min reaction time privigen®, Intratect®, Kiovig®, and Gammunex® had no approx. SC200 or approx. IC50. Gammagard® had an approx. IC50 of 4 g/l added IgG (Figure 6). With EDTA-plasma analyzed at 59 min the approx. IC50 values were 1 g/l Gammagard, 5 g/l Gammunex, 5 g/l privigen, or 6 g/l Kiovig. Intratect® had neither an IC50 nor a SC200 (Figure 7).

At 70 min reaction time all IgG drugs had no approx. SC200 or approx. IC50 on blood ROS generation (Figure 8).

At 90 min reaction time Gammagard®, Kiovig®, or Gammunex® had an approx. IC50 of 2, 5, or 9 g/l added IgG; privigen® and Intratect® had neiter IC50 nor SC200 (Figure 9).

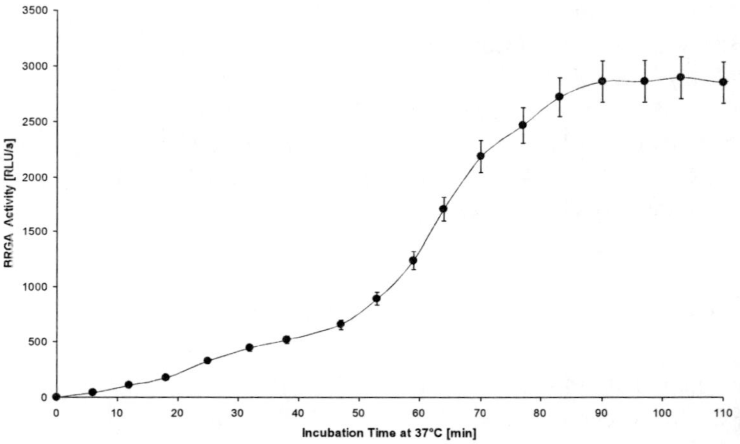

Figure 1. Kinetic of blood ROS generation increase in citrated blood. 125 µl HBSS were incubated in 10-fold with 40 µl 0.9% NaCl, 10 µl normal citrated blood (stored for 1d at 23°C), 10 µl 0.26 mM luminol (final), and 10 µl 1.9 µg/ml (final) zymosan A in black Brand®781608 polystyrene microwells.

Discussion

The BRGA for citrated blood imitates better the *in vivo* situation than that for EDTA blood. Short incubation times (here 25-38 min) demonstrate the direct action of IgG preparations on blood neutrophils. At prolonged incubation times the ROS generation stimulatory action of IgG, especially of complement fixing IgG complexes (or of denatured IgG) that stimulate neutrophils via FcgammaR2 or FcgammaR3 receptors [12-29], were lost and all IgG drugs became more or less inhibitory towards blood ROS generation.

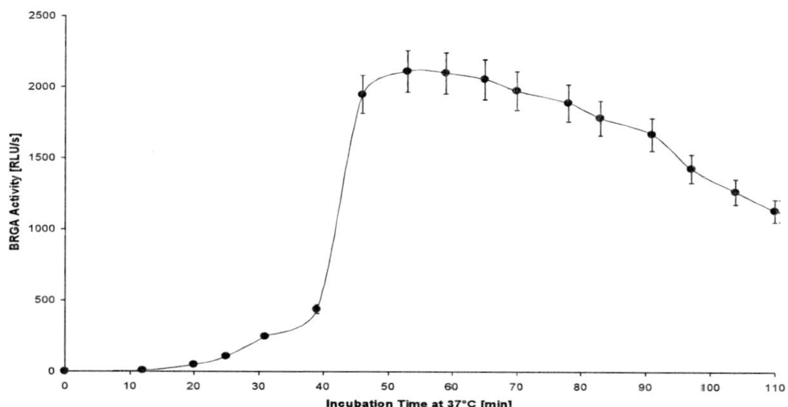

Figure 2. Kinetic of blood ROS generation in EDTA blood. 125 µl HBSS (modified without phenol red) were incubated in 10-fold with 40 µl 0.9% NaCl, 10 µl normal EDTA blood (stored for 1d at 23°C), 10 µl 0.26 mM luminol (final), and 10 µl 1.9 µg/ml (final) zymosan A in black Brand®781608 polystyrene microwells.

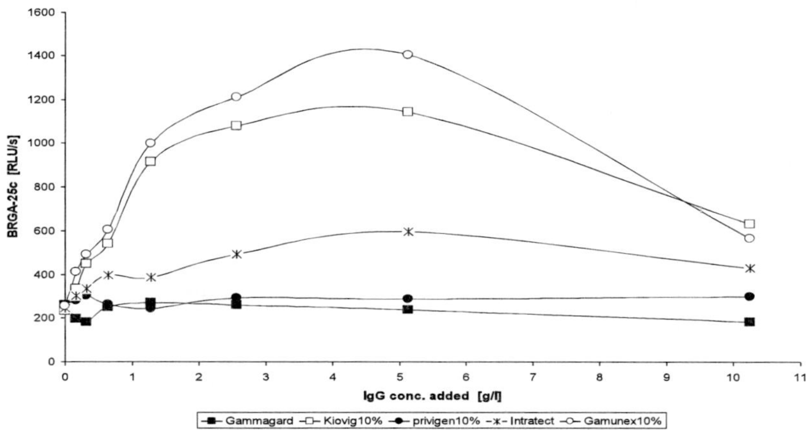

Figure 3. Blood ROS generation in presence of human IgG drugs. 40 µl 0-10.3 g/l (final) privigen® (●), Gammagard® (■), Intratect® (*), Kiovig® (□), Gammunex® (O) in 0.9% NaCl (all IgG drugs in -30°C frozen/ 23°C thawed aliquots of 50 g/l) in black Brand®781608 high quality F-wells were incubated in triplicate with 125 µl HBSS (modified without phenol red), 10 µl citrated blood, 10 µl 0.26 mM (final) luminol, and 10 µl 1.9 µg/ml (final) zymosan A. The plates were measured at 25 min (37°C) in a photons-multiplyer microtiter plate luminometer (LUmo).

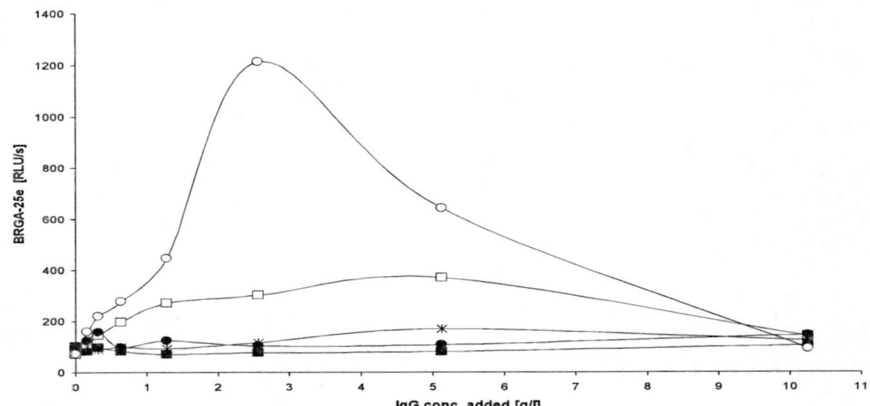

Figure 4. Blood ROS generation in presence of human IgG drugs. 40 µl 0-10.3 g/l (final) privigen® (●), Gammagard® (■), Intratect® (*), Kiovig® (□), Gammunex® (O) in 0.9% NaCl in black Brand®781608 high quality polystyrene microwells were incubated in triplicate with 125 µl Hanks' balanced salt solution (HBSS; modified without phenol red), 10 µl EDTA blood, 10 µl 0.26 mM (final) luminol, and 10 µl 1.9 µg/ml (final) zymosan A.

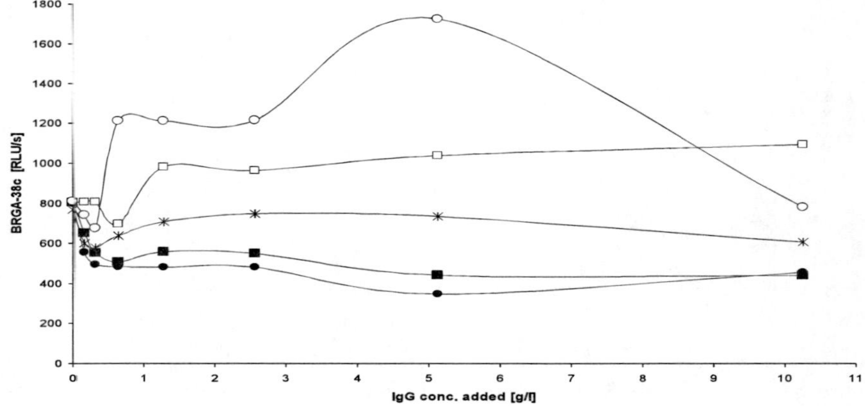

Figure 5. Blood ROS generation in presence of human IgG drugs. 40 µl 0-10.3 g/l (final) privigen® (●), Gammagard® (■), Intratect® (*), Kiovig® (□), Gammunex® (O) in 0.9% NaCl in black Brand®781608 high quality polystyrene microwells were incubated in triplicate with 125 µl Hanks' balanced salt solution (HBSS; modified without phenol red), 10 µl citrated blood, 10 µl 0.26 mM (final) luminol, and 10 µl 1.9 µg/ml (final) zymosan A.

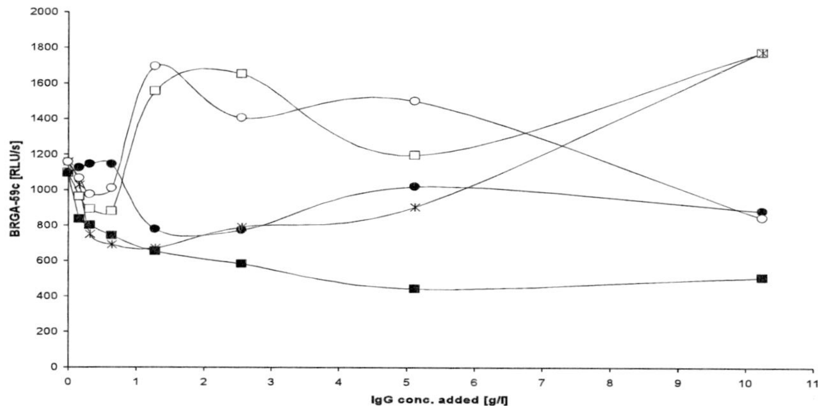

Figure 6. Blood ROS generation in presence of human IgG drugs. 40 µl 0-10.3 g/l (final) privigen® (●), Gammagard® (■), Intratect® (*), Kiovig® (□), Gammunex® (O) in 0.9% NaCl in black Brand®781608 high quality polystyrene microwells were incubated in triplicate with 125 µl HBSS, 10 µl citrated blood, 10 µl 0.26 mM (final) luminol, and 10 µl 1.9 µg/ml (final) zymosan A. The plates were measured at 59 min (37°C) in a photons-multiplyer microtiter plate luminometer.

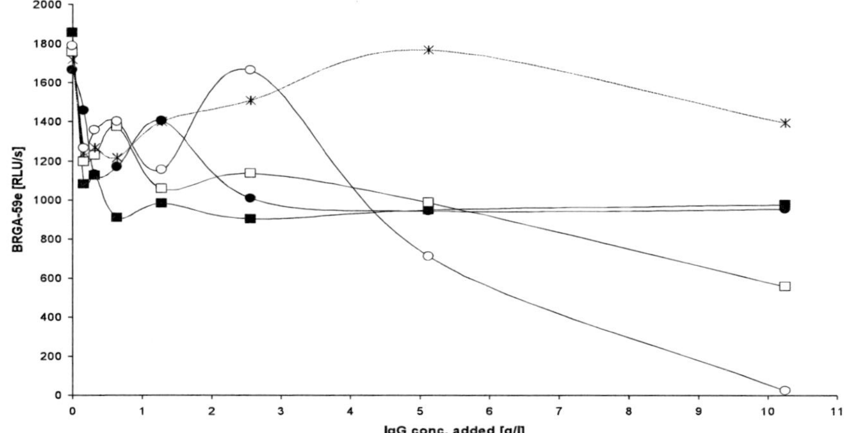

Figure 7. Blood ROS generation in presence of human IgG drugs. 40 µl 0-10.3 g/l (final) privigen® (●), Gammagard® (■), Intratect® (*), Kiovig® (□), Gammunex® (O) in 0.9% NaCl in black Brand®781608 high quality polystyrene microwells were incubated in triplicate with 125 µl Hanks′ balanced salt solution (HBSS; modified without phenol red), 10 µl EDTA blood, 10 µl 0.26 mM (final) luminol, and 10 µl 1.9 µg/ml (final) zymosan A. The plates were measured at 59 min (37°C) in a photons-multiplyer microtiter plate luminometer (LUmo).

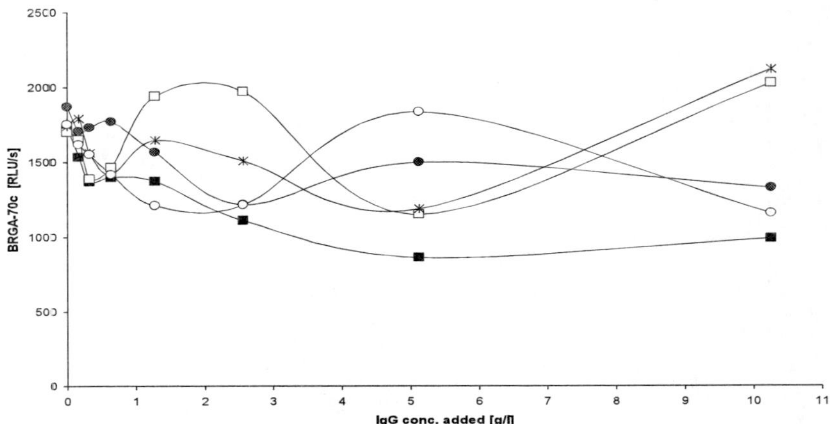

Figure 8. Blood ROS generation in presence of human IgG drugs. 40 µl 0-10.3 g/l (final) privigen® (●), Gammagard® (■), Intratect® (*), Kiovig® (☐), Gammunex® (O) in 0.9% NaCl in black Brand®781608 high quality polystyrene microwells were incubated in triplicate with 125 µl Hanks′ balanced salt solution (HBSS; modified without phenol red), 10 µl citrated blood, 10 µl 0.26 mM (final) luminol, and 10 µl 1.9 µg/ml (final) zymosan A. The plates were measured at 70 min (37°C) in a photons-multiplyer microtiter plate luminometer (LUmo).

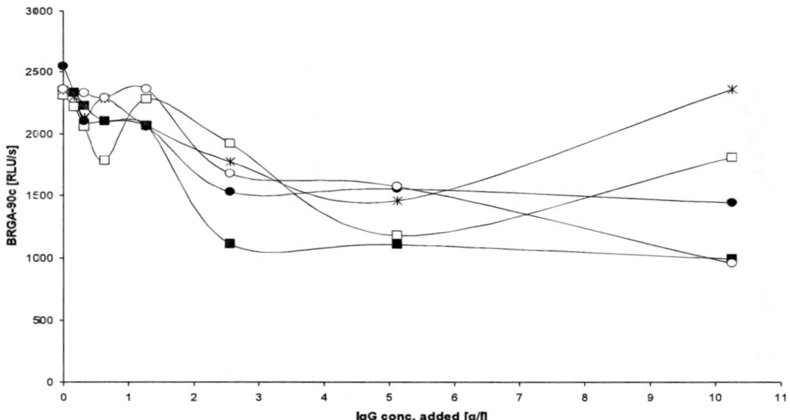

Figure 9. Blood ROS generation in presence of human IgG drugs. 40 µl 0-10.3 g/l (final) privigen® (●), Gammagard® (■), Intratect® (*), Kiovig® (☐), Gammunex® (O) in 0.9% NaCl in black Brand®781608 high quality polystyrene microwells were incubated in triplicate with 125 µl Hanks′ balanced salt solution (HBSS; modified without phenol red), 10 µl citrated blood, 10 µl 0.26 mM (final) luminol, and 10 µl 1.9 µg/ml (final) zymosan A. The plates were measured at 90 min (37°C) in a photons-multiplyer microtiter plate luminometer (LUmo).

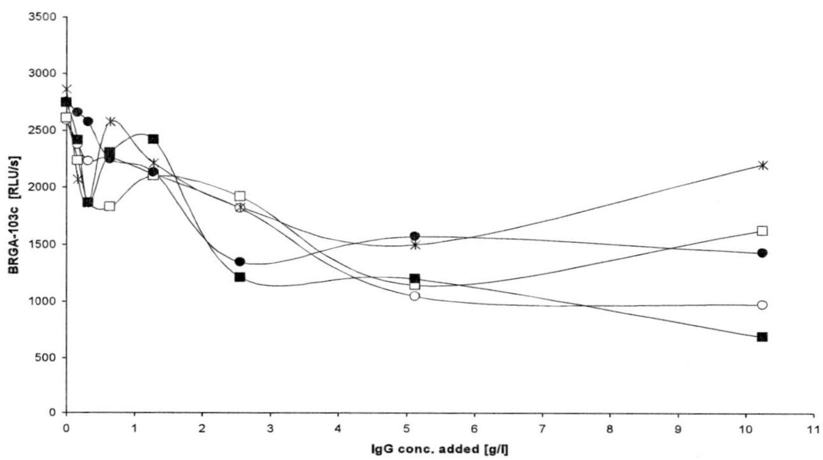

Figure 10. Blood ROS generation in presence of human IgG drugs. 40 µl 0-10.3 g/l (final) privigen®-Behring (●), Gammagard®-Baxter (■), Intratect®-Biotest (*), Kiovig®-Baxter (□), Gammunex®-Talecris (O) in 0.9% NaCl in black Brand®781608 high quality polystyrene microwells were incubated in triplicate with 125 µl Hanks′ balanced salt solution (HBSS; modified without phenol red), 10 µl citrated blood, 10 µl 0.26 mM (final) luminol, and 10 µl 1.9 µg/ml (final) zymosan A.

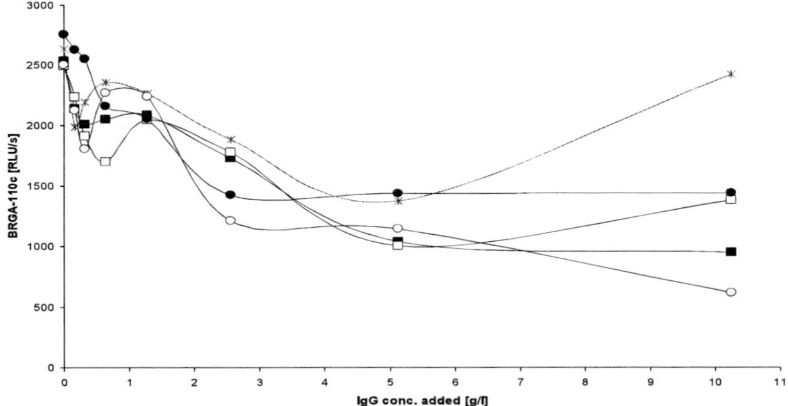

Figure 11. Blood ROS generation in presence of human IgG drugs. 40 µl 0-10.3 g/l (final) privigen® (●), Gammagard® (■), Intratect® (*), Kiovig® (□), Gammunex® (O) in 0.9% NaCl in black Brand®781608 high quality polystyrene microwells were incubated in triplicate with 125 µl HBSS, 10 µl citrated blood, 10 µl 0.26 mM (final) luminol, and 10 µl 1.9 µg/ml (final) zymosan A. The plates were measured at 110 min (37°C) in a photons-multiplyer microtiter plate luminometer.

At 103 min reaction time only privigen® and Intratect® had neither IC50 nor SC200. The approx. IC50 of Gammagard® was 3 g/l, that of Kiovig® and Gammunex® 5 g/l added IgG (Figure 10).

At 110 min reaction time again only privigen® and Intratect® had neither IC50 nor SC200. The approx. IC50 of Gammunex® was 3 g/l, that of Kiovig® and Gammagard® 5 g/l added IgG (Figure 11).

Patients that need IVIG require infusion of an IgG drug of high quality. The IgG drug should

1) function as IgG
2) not stimulate blood ROS generation (hyper-stimulation could result in systemic inflammation [30-33] with decreased on endothelium patrolling neutrophils [34])
3) not inhibit blood ROS generation too strongly
4) not act as an activator of AM-coagulation (coagulation triggered by altered matrix [35]).

Point 1 cannot be judged in the hemostasis laboratory. Here special microbiological techniques or clinical correlates are necessary to obtain valid results on the real function of IgG. Points 2 and 3 have been analyzed in the present work, point 4 has previously been analyzed [35]. From all the data on points 2-4 together it is suggested that privigen® could be the best *i.v.* human IgG drug currently on the medical market. Standardized large studies on blood ROS generation modulation by different IgG drugs are required.

The BRGA technique is a powerful procedure to globally test the important cellular H_2O_2 generation [36]. The few specific triggers acting via specific proteinous receptors such as CD11b/18 [37] or FcR [13,38] or via activation of protein kinase C [39] and the many different primers of blood ROS generation [40] can be studied in a matrix that is very similar to flowing blood. All these stimulators or inactivators [41] modulate the cellular assembly of the NADPH-oxidase, consisting of the 4 cytoplasmic enzyme parts $p67^{phox}$ (bearing the NADPH), $p40^{phox}$ (the $p67^{phox}$ stabilizer and phosphoinositide binder), $p47^{phox}$, GTP-Rac (the enzyme regulator linked to Ca^{2+} ion currents) that associate with the main membranous enzyme part $gp91^{phox}$ (bearing the NADPH oxidizing cytochrome b558 with its FAD and two hemes) [38]. The fully activated NADPH-oxidase generates large amounts of the mother ROS H_2O_2 [42,43] that gives rise to the two daughter ROS hydroxyl-radical (•OH) and (together with secreted myeloperoxidase) non-radicalic excited singlet oxygen ($^1O_2$*) [44-46], the precursor of excited carbonyls (C=O*) that emit photons at about 400 nm. Thus, $^1O_2$* is both destructive weapon against thrombi, bacteria, fungi, virus, malaria parasites, and a cell signal [47-53].

There are exciting times ahead of all medical researchers that focus on the biochemistry, pathobiochemistry, physiology, pathophysiology, pharmacology, toxicology, histology, histopathology of $^1O_2$* [54-56].

REFERENCES

[1] Aukrust P, Müller F, Svenson M, Nordøy I, Bendtzen K, Frøland SS. Administration of intravenous immunoglobulin (IVIG) in vivo--down-regulatory effects on the IL-1 system. *Clin Exp Immunol.* 1999; 115: 136-43.

[2] Klaesson S, Ringdén O, Markling L, Remberger M, Lundkvist I. Immune modulatory effects of immunoglobulins on cell-mediated immune responses in vitro. *Scand J Immunol.* 1993; 38: 477-84.

[3] Sibéril S, Elluru S, Negi VS, Ephrem A, Misra N, Delignat S, Bayary J, Lacroix-Desmazes S, Kazatchkine MD, Kaveri SV. Intravenous immunoglobulin in autoimmune and inflammatory diseases: more than mere transfer of antibodies. *Transfus Apher Sci.* 2007; 37: 103-7.

[4] Semple JW, Kim M, Hou J, McVey M, Lee YJ, Tabuchi A, Kuebler WM, Chai ZW, Lazarus AH. Intravenous immunoglobulin prevents murine antibody-mediated acute lung injury at the level of neutrophil reactive oxygen species (ROS) production. *PLoS One.* 2012; 7: e31357.

[5] Aukrust P, Müller F, Frøland SS. Enhanced generation of reactive oxygen species in monocytes from patients with common variable immunodeficiency. *Clin Exp Immunol.* 1994; 97: 232-8.

[6] Reagan WJ, Scott-Moncrieff C, Christian J, Snyder P, Kelly K, Glickman L. Effects of human intravenous immunoglobulin on canine monocytes and lymphocytes. *Am J Vet Res.* 1998; 59: 1568-74.

[7] Stief T.W. Neutrophil granulocytes in hemostasis *Hemostasis Laboratory* 2009; 2: 269-89.

[8] Ano Y, Sakudo A, Onodera T. Role of microglia in oxidative toxicity associated with encephalomyocarditis virus infection in the central nervous system. *Int J Mol Sci.* 2012; 13: 7365-74.

[9] Savina A, Vargas P, Guermonprez P, Lennon AM, Amigorena S. Measuring pH, ROS production, maturation, and degradation in

dendritic cell phagosomes using cytofluorometry-based assays. *Methods Mol Biol*. 2010; 595: 383-402.

[10] Stief T. The routine blood ROS generation assay (BRGA) triggered by typical septic concentrations of zymosan A. *Hemost Lab.* 2013; 6: 89-98.

[11] Stief TW. Reactive oxygen species generation in diluted whole blood anticoagulated by citrate, EDTA, or heparin. *Hemostasis Laboratory* 2013; 6: (issue 2-3)

[12] Hung SL, Chiang HH, Wu CY, Hsu MJ, Chen YT. Effects of herpes simplex virus type 1 infection on immune functions of human neutrophils. *J Periodontal Res*. 2012; 47: 635-44.

[13] Higurashi S, Machino Y, Suzuki E, Suzuki M, Kohroki J, Masuho Y. Both the Fab and Fc domains of IgG are essential for ROS emission from TNF-α-primed neutrophils by IVIG. *Biochem Biophys Res Commun*. 2012; 417: 794-9.

[14] Jarius S, Eichhorn P, Albert MH, Wagenpfeil S, Wick M, Belohradsky BH, Hohlfeld R, Jenne DE, Voltz R. Intravenous immunoglobulins contain naturally occurring antibodies that mimic antineutrophil cytoplasmic antibodies and activate neutrophils in a TNFalpha-dependent and Fc-receptor-independent way. *Blood.* 2007; 109: 4376-82.

[15] Rasheed Z, Al-Shobaili HA, Alzolibani AA, Ismail Khan M, Tariq Ayub M, Khan MI, Rasheed N. Immunological functions of oxidized human immunoglobulin G in type 1 diabetes mellitus: its potential role in diabetic smokers as a biomarker of elevated oxidative stress. *Dis Markers*. 2011; 31: 47-54.

[16] Al-Shobaili HA, Al Robaee AA, Alzolibani A, Khan MI, Rasheed Z. Hydroxyl radical modification of immunoglobulin generated cross-reactive antibodies: its potential role in systemic lupus erythematosus. *Clin Med Insights Arthritis Musculoskelet Disord*. 2011; 4: 11-9.

[17] Kulkarni S, Sitaru C, Jakus Z, Anderson KE, Damoulakis G, Davidson K, Hirose M, Juss J, Oxley D, Chessa TA, Ramadani F, Guillou H, Segonds-Pichon A, Fritsch A, Jarvis GE, Okkenhaug K, Ludwig R, Zillikens D, Mocsai A, Vanhaesebroeck B, Stephens LR, Hawkins PT. PI3Kβ plays a critical role in neutrophil activation by immune complexes. *Sci Signal*. 2011; 4: ra23.

[18] Ottonello L, Frumento G, Arduino N, Dapino P, Tortolina G, Dallegri F. Immune complex stimulation of neutrophil apoptosis: investigating the

involvement of oxidative and nonoxidative pathways. *Free Radic Biol Med.* 2001; 30: 161-9.

[19] Tasneem S, Ali R. Binding of SLE autoantibodies to native poly(I), ROS-poly(I) and native DNA: a comparative study. *J Autoimmun.* 2001;17:199-205.

[20] Abdi S, Ali A. Role of ROS modified human DNA in the pathogenesis and etiology of cancer. *Cancer Lett.* 1999; 142: 1-9.

[21] Liu L, Elwing H, Karlsson A, Nimeri G, Dahlgren C. Surface-related triggering of the neutrophil respiratory burst. Characterization of the response induced by IgG adsorbed to hydrophilic and hydrophobic glass surfaces. *Clin Exp Immunol.* 1997; 109: 204-10.

[22] Andersson U, Björk L, Skansén-Saphir U, Andersson J. Pooled human IgG modulates cytokine production in lymphocytes and monocytes. *Immunol Rev.* 1994; 139: 21-42.

[23] Griffiths HR, Lunec J, Jefferis R, Blake DR, Willson RL. A study of ROS induced denaturation of IgG3 using monoclonal antibodies; implications for inflammatory joint disease. *Basic Life Sci.* 1988; 49: 361-4.

[24] Djoumerska-Alexieva IK, Dimitrov JD, Nacheva J, Kaveri SV, Vassilev TL. Protein destabilizing agents induce polyreactivity and enhanced immunomodulatory activity in IVIg preparations. *Autoimmunity.* 2009; 42: 365-7.

[25] Rasheed Z, Khan MW, Ali R. Hydroxyl radical modification of human serum albumin generated cross reactive antibodies. *Autoimmunity.* 2006; 39: 479-88.

[26] Dimitrov JD, Vassilev TL, Andre S, Kaveri SV, Lacroix-Desmazes S. Functional variability of antibodies upon oxidative processes. *Autoimmun Rev.* 2008; 7:574-8.

[27] Alam K, Moinuddin, Jabeen S. Immunogenicity of mitochondrial DNA modified by hydroxyl radical. *Cell Immunol.* 2007; 247: 12-7.

[28] Chedraoui-Silva S, Mantovani B. The role of complement in the modulation by fluid-phase IgG of the production of reactive oxygen species by polymorphonuclear leukocytes stimulated with IgG immune complexes. *Braz J Med Biol Res.* 2003; 36: 1665-72.

[29] Al Arfaj AS, Chowdhary AR, Khalil N, Ali R. Immunogenicity of singlet oxygen modified human DNA: implications for anti-DNA antibodies in systemic lupus erythematosus. *Clin Immunol.* 2007; 124: 83-9.

[30] Martins PS, Kallas EG, Neto MC, Dalboni MA, Blecher S, Salomão R. Upregulation of reactive oxygen species generation and phagocytosis, and increased apoptosis in human neutrophils during severe sepsis and septic shock. *Shock.* 2003; 20: 208-12.

[31] Alves-Filho JC, de Freitas A, Spiller F, Souto FO, Cunha FQ. The role of neutrophils in severe sepsis. *Shock.* 2008; 30 Suppl 1: 3-9.

[32] Matthews JB, Chen FM, Milward MR, Wright HJ, Carter K,McDonagh A, Chapple IL. Effect of nicotine, cotinine and cigarette smoke extract on the neutrophil respiratory burst. *J Clin Periodontol.* 2011; 38: 208-18.

[33] Fairhurst AM, Wallace PK, Jawad AS, Goulding NJ. Rheumatoid peripheral blood phagocytes are primed for activation but have impaired Fc-mediated generation of reactive oxygen species. *Arthritis Res Ther.* 2007; 9: R29.

[34] Russwurm S, Vickers J, Meier-Hellmann A, Spangenberg P, Bredle D, Reinhart K, Lösche W. Platelet and leukocyte activation correlate with the severity of septic organ dysfunction. *Shock.* 2002; 17: 263-8.

[35] Stief TW. Thrombin generation by therapeutic immunoglobulins. In: *Thrombin: function and pathophysiology.* Stief T (ed.), Nova Science Publishers, New York, 2012, pp. 47-58.

[36] Attar H, Bedard K, Migliavacca E, Gagnebin M, Dupré Y, Descombes P, Borel C, Deutsch S, Prokisch H, Meitinger T, Mehta D, Wichmann E, Delabar JM, Dermitzakis ET, Krause KH, Antonarakis SE. Extensive natural variation for cellular hydrogen peroxide release is genetically controlled. *PLoS One.* 2012; 7: e43566.

[37] Goodridge HS, Wolf AJ, Underhill DM. Beta-glucan recognition by the innate immune system. *Immunol Rev.* 2009; 230: 38-50.

[38] Anderson KE, Chessa TA, Davidson K, Henderson RB, Walker S, Tolmachova T, Grys K, Rausch O, Seabra MC,Tybulewicz VL, Stephens LR, Hawkins PT. PtdIns3P and Rac direct the assembly of the NADPH oxidase on a novel, pre-phagosomal compartment during FcR-mediated phagocytosis in primary mouse neutrophils. *Blood* 2010; 116: 4978-89.

[39] Morigi M, Macconi D, Zoja C, Donadelli R, Buelli S, Zanchi C, Ghilardi M, Remuzzi G. Protein overload-induced NF-kappaB activation in proximal tubular cells requires H_2O_2 through a PKC-dependent pathway. *J Am Soc Nephrol.* 2002; 13: 1179-89.

[40] Stief T. Light quants of low wave length prime blood neutrophils for ROS generation. *Hemost Lab* 2013; 6 (issue 4).

[41] Shawcross DL, Wright GA, Stadlbauer V, Hodges SJ, Davies NA, Wheeler-Jones C, Pitsillides AA, Jalan R. Ammonia impairs neutrophil phagocytic function in liver disease. *Hepatology*. 2008; 48: 1202-12.

[42] Nathan CF. Neutrophil activation on biological surfaces. Massive secretion of hydrogen peroxide in response to products of macrophages and lymphocytes. *J Clin Invest*. 1987; 80: 1550-60.

[43] Seguchi H, Kobayashi T. Study of NADPH oxidase-activated sites in human neutrophils. *J Electron Microsc*. 2002; 51: 87-91.

[44] Rosen H, Klebanoff SJ. Formation of singlet oxygen by the myeloperoxidase-mediated antimicrobial system. *J Biol Chem*. 1977; 252: 4803-10.

[45] Kiryu C, Makiuchi M, Miyazaki J, Fujinaga T, Kakinuma K. Physiological production of singlet molecular oxygen in the myeloperoxidase-H_2O_2-chloride system. *FEBS Lett*. 1999; 443: 154-8.

[46] Maghzal G, Krause KH, Stocker R, Jaquet V. Detection of reactive oxygen species derived from the family of NOX (NADPH oxidases). *Free Radic Biol Med*. 2012; 53: 1903-18.

[47] Briviba K, Klotz LO, Sies H. Toxic and signaling effects of photochemically or chemically generated singlet oxygen in biological systems. *Biol Chem*. 1997; 378: 1259-65.

[48] Stief TW, Fareed J. The antithrombotic factor singlet oxygen/light ($^1O_2$/hv). *Clin Appl Thrombosis/Hemostasis* 2000; 6: 22-30.

[49] Stief TW. The blood fibrinolysis / deep - sea analogy: a hypothesis on the cell signals singlet oxygen / photons as natural antithrombotics. *Thromb Res*. 2000; 99: 1-20.

[50] Stief TW. The physiology and pharmacology of singlet oxygen. *Med Hypoth*. 2003; 60: 567-72.

[51] Stief TW. Regulation of hemostasis by singlet-oxygen ($^1\Delta O_2$). *Curr Vasc Pharmacol* 2004; 2: 357-62.

[52] Joos C, Marrama L, Polson HE, Corre S, Diatta AM, Diouf B, Trape JF, Tall A, Longacre S, Perraut R. Clinical protection from falciparum malaria correlates with neutrophil respiratory bursts induced by merozoites opsonized with human serum antibodies. *PLoS One*. 2010; 5: e9871.

[53] Legrand-Poels S, Schoonbroodt S, Matroule JY, Piette J. Nf-kappa B: an important transcription factor in photobiology. *J Photochem Photobiol B*. 1998; 45: 1-8.

[54] Couser WG, Pippin JW, Shankland SJ. Complement (C5b-9) induces DNA synthesis in rat mesangial cells in vitro. *Kidney Int.* 2001; 59: 905-12.

[55] Tezel G, Yang X, Luo C, Peng Y, Sun SL, Sun D. Mechanisms of immune system activation in glaucoma: oxidative stress-stimulated antigen presentation by the retina and optic nerve head glia. *Invest Ophthalmol Vis Sci.* 2007; 48: 705-14.

[56] Fortin CF, McDonald PP, Lesur O, Fülöp T Jr. Aging and neutrophils: there is still much to do. *Rejuvenation Res.* 2008; 11: 873-82.

Chapter 7

Naproxen Is Only a Mild Inhibitor of Neutrophils

Abstract

Background: The non-steroidal anti-inflammatory drug (NSAID) naproxen is less thrombogenic than other NSAIDs. Naproxen is an about 20-fold weaker contact trigger of hemostasis than ibuprofen. Has the drug naproxen a tolerable action on blood neutrophils to be generally recommended as standard NSAID. The main function of blood neutrophils is the generation of reactive oxygen species (ROS), detectable by the blood ROS generation assay (BRGA).

Material and Methods: 10 µl samples of individual normal citrated blood were added to 125 µl Hanks′ Balanced Salt Solution and 40 µl 0-1.28 g/l naproxen in 0.9% NaCl (0-262 mg/l final) in high quality polystyrene black microwells (Brand®781608). Immediately (for the main assay version BRGA) and after 60 min at 37°C (for the pre-incubation assay version BRGA-60-) 10 µl 5 mM(0.26 mM final) luminol and 10 µl 36 µg/ml (1.9 µg/ml final) zymosan A were added. The light emission per well was determined by a photons-multiplied microtiter plate luminomenter (LUmo) with an integration time of 0.5s. The approximate 50% inhibitory concentrations (approx. IC50) were determined in the increasing part of the blood ROS generation curve.

Results and Discussion: The approx. IC50 in the initial phase of the BRGA was 0.5-2.5 mg/l naproxen. An approx. IC75 did not appear. In BRGA-60-41 (BRGA with 60 min pre-incubation to activate intra-cellular enzyme systems such as cytochrome P450 (CYP450) and 41 min main incubation) the approx. IC50 was 3-8 mg/l (2 of 3 samples) naproxen without an approx. IC75. The BRGA in its initial phase reflects better the neutrophil inhibitory action of naproxen than later BRGA reaction times or than the any reaction time of the BRGA-60-, where naproxen is increasingly metabolized by cellular CYP450 into a stimulator of blood ROS generation. Compared with other anti-neutrophil ROS drugs such as diclofenac or acetaminophen naproxen is only a mild inhibitor of blood ROS generation. Therefore, naproxen can be prescribed without fearing a too dramatic decrease of the neutrophil′s function. If there is medical need for suppression of the neutrophils then naproxen should be combined with the weak contact phase - trigger acetaminophen.

Keywords: Naproxen, NSAID, neutrophils, ROS, BRGA

INTRODUCTION

Naproxen is a non-steroidal anti-inflammatory drug (NSAID) that is less thrombogenic than other NSAIDs [1-4]. The ultra-specific, ultra-sensitive plasmatic thrombin generation assay RECA established about 20-fold elevated approx. SC200 values of naproxen when compared with ibuprofen, i.e. naproxen is an about 20-fold weaker contact trigger of hemostasis than ibuprofen [5]. If the drug naproxen has a tolerable action on blood neutrophils it could generally be generally as standard NSAID. The main function of blood neutrophils is the generation of reactive oxygen species (ROS) that is detectable by the blood ROS generation assay (BRGA) [6].

MATERIAL AND METHODS

10 µl samples of individual normal citrated blood (polypropylene monovettes for 9 parts of blood pre-filled with 1 part of 106 mM Na_3-citrate, pH 7.4; Sarstedt, Nümbrecht, Germany) were added in triplicate after written informed consent to 125 µl Hanks′ Balanced Salt Solution (Sigma, Deisenhofen, Germany) and 40 µl 0-1.28 g/l naproxen (Sigma-Fluka) in 0.9% NaCl (0-262 mg/l final) in high quality polystyrene black microwells (Brand, Wertheim, Germany; article nr. 781608). Immediately (for the main assay version BRGA) and after 60 min at 37°C (for the pre-incubation assay version BRGA-60-) 10 µl 5 mM (0.26 mM final) luminol sodium salt (Sigma) and 10 µl 36 µg/ml (1.9 µg/ml final) zymosan A (Sigma) were added. The light emission per well was determined by a photons-multiplied microtiter plate luminomenter (LUmo; anthos-Autobio, Krefeld, Germany) with an integration

time of 0.5s. The mean values of the triplicate determinations were calculated, the intra-assay coefficients of variation were less than 10%. The approximate 50% inhibitory concentrations (approx. IC50) were determined in the increasing part of the blood ROS generation curve.

Results and Discussion

The ROS maxima of about 5000-6000 RLU/s were reached after 70 min for N1, N2 and of about 900 RLU/s after 54 min for N3. At 206 min (N1,N2) or 122 min (N3) additional smaller ROS maxima appeared, the onset of cellular fibrinolysis [7] (Figure 1).

The ROS maxima of about 9000-11000 RLU/s (N1, N2) or about 1200 RLU/s (N3) were reached after about 70 min (Figure 2).

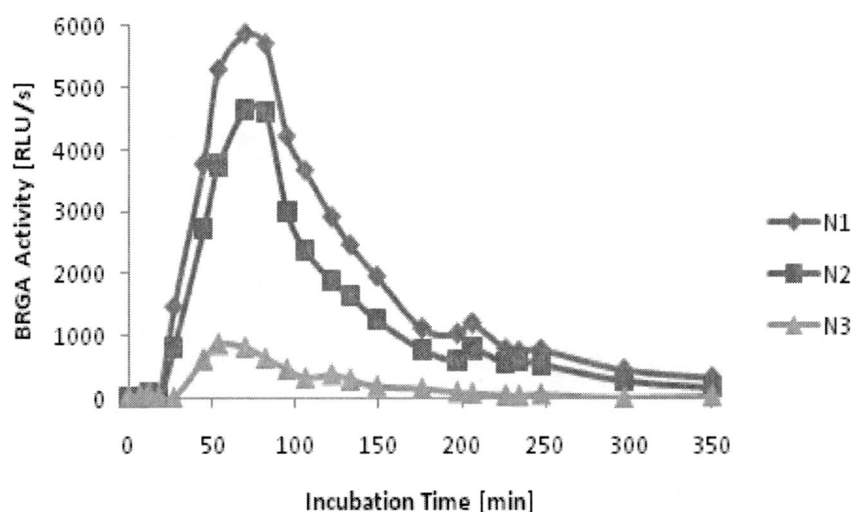

Figure 1. BRGA kinetic in normal citrated blood. Normal citrated blood samples (N1 and N2, stored for 1d at 23°C and N3, stored up to 2d at 23°C) were analyzed in the BRGA.

The approx. IC50 for samples N1 and N2 in BRGA-18 (BRGA with 18 min reaction time) was 0.5 mg/l naproxen. An approx. IC75 did not appear (Figure 3). An approx. IC50 did not appear in BRGA-45 (Figure 4).

N3 analyzed in BRGA-45 had an approx. IC50 of 2.5 mg/l naproxen. An approx. IC75 did not appear (Figure 5).

In BRGA-60-24 (BRGA with 60 min pre-incubation to activate intra-cellular enzyme systems such as cytochrome P450 (CYP450) [8-10] and 24 min main incubation) the approx. IC50 was 3 mg/l naproxen for N1 and 130 mg/l naproxen for N2. An approx. IC75 did not appear (Figure 6).

In BRGA-60-41 the approx. IC50 was 4 mg/l naproxen for N1 and 50 mg/l naproxen for N2, with an approx. IC75 of 200 mg/l naproxen. For N3 there appeared only an approx. SC200 of 8 mg/l naproxen (Figure 7).

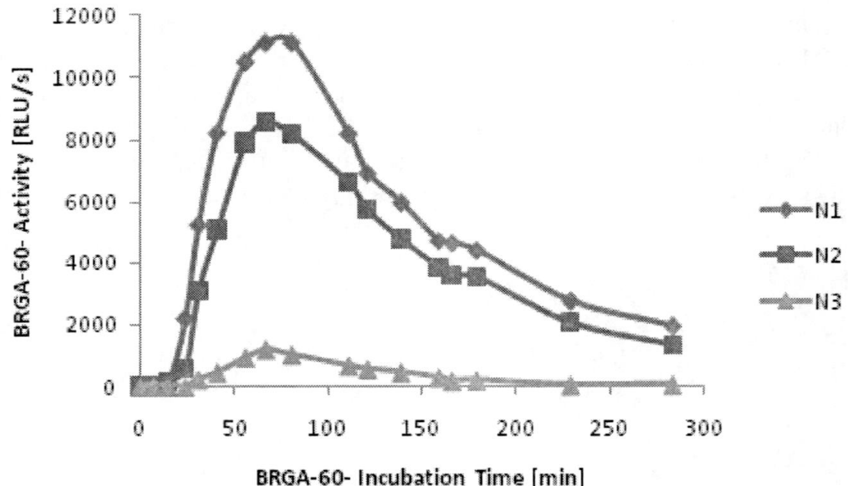

Figure 2. BRGA-60- kinetic in normal citrated blood. Normal citrated blood samples (N1 and N2, stored for 1d at 23°C and N3, stored up to 2d at 23°C) were analyzed in the BRGA-60-, i.e. the BRGA version with a 60 min blood in HBSS pre-incubation time prior to addition of luminol and 1.9 µg/ml zymosan A that starts the main incubation.

Thus, the BRGA in its initial phase reflects better the neutrophil inhibitory action of naproxen than later BRGA reaction times or than the any reaction time of the BRGA-60-, where naproxen is increasingly metabolized by cellular CYP450 into a stimulator of blood ROS generation. Compared with other anti-neutrophil ROS drugs such as diclofenac or acetaminophen naproxen is only a mild inhibitor of blood ROS generation [11,12]. There was no suppression of blood ROS generation below the 25% line (seen in IC75). Therefore, naproxen can be prescribed without fearing a too dramatic decrease of the neutrophil´s function [13-18]. If there is medical need for suppression of the neutrophils then a naproxen should be combined with the weak altered matrix- trigger acetaminophen [19-21].

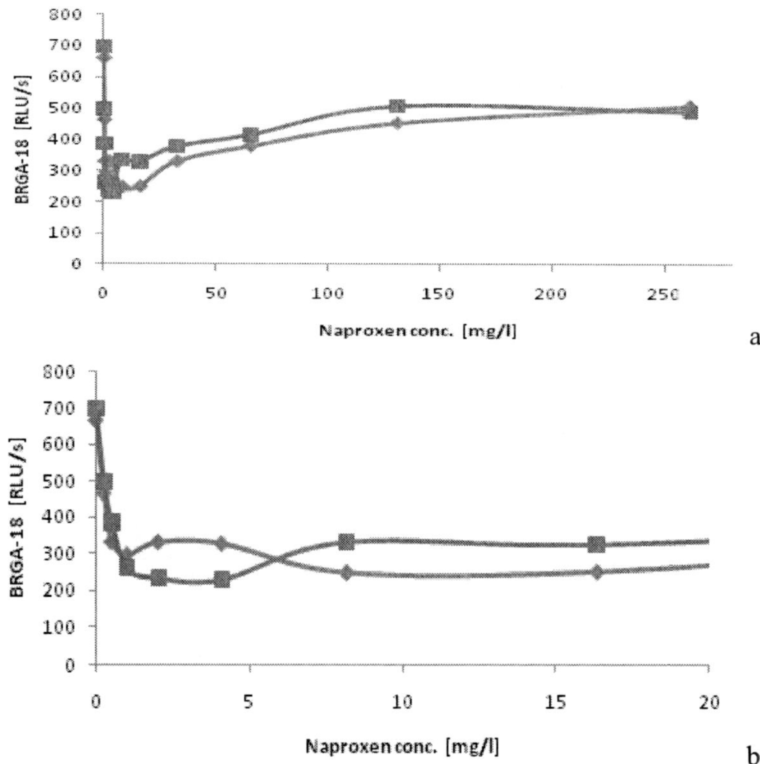

Figure 3. Inhibition of blood ROS generation by naproxen in BRGA. The BRGA-18 (BRGA with 18 min reaction time) was performed for blood samples N1 and N2. The approx. IC50 was 0.5 mg/l naproxen. An approx. IC75 did not appear.

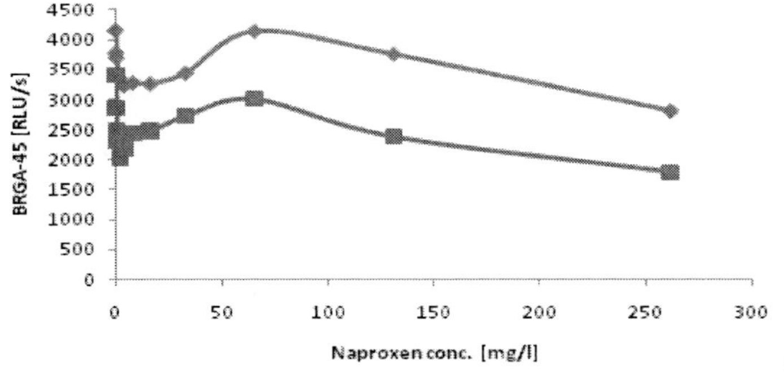

Figure 4. (Continued)

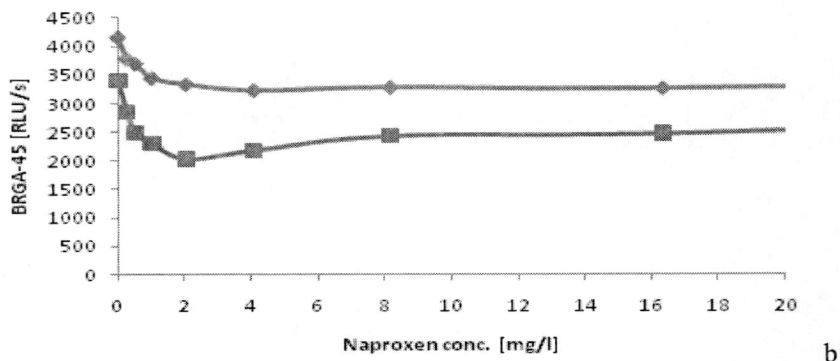

Figure 4. Inhibition of blood ROS generation by naproxen in BRGA. The BRGA-45 (BRGA with 45 min reaction time) was performed for blood samples N1 and N2. An approx. IC50 did not appear.

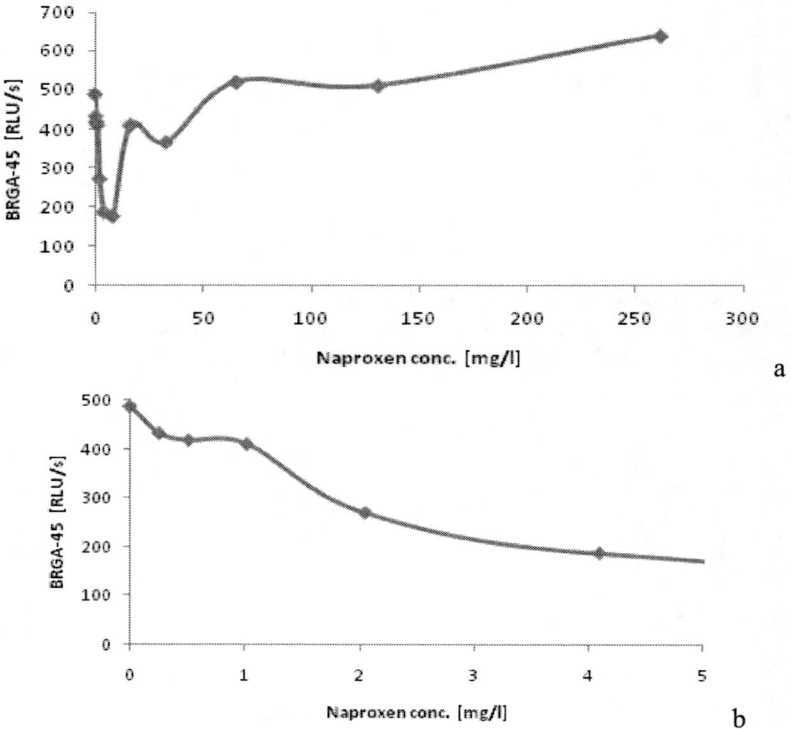

Figure 5. Inhibition of blood ROS generation by naproxen in BRGA. The BRGA-45 (BRGA with 45 min reaction time) was performed for blood sample N3. The approx. IC50 was 2.5 mg/l naproxen. An approx. IC75 did not appear.

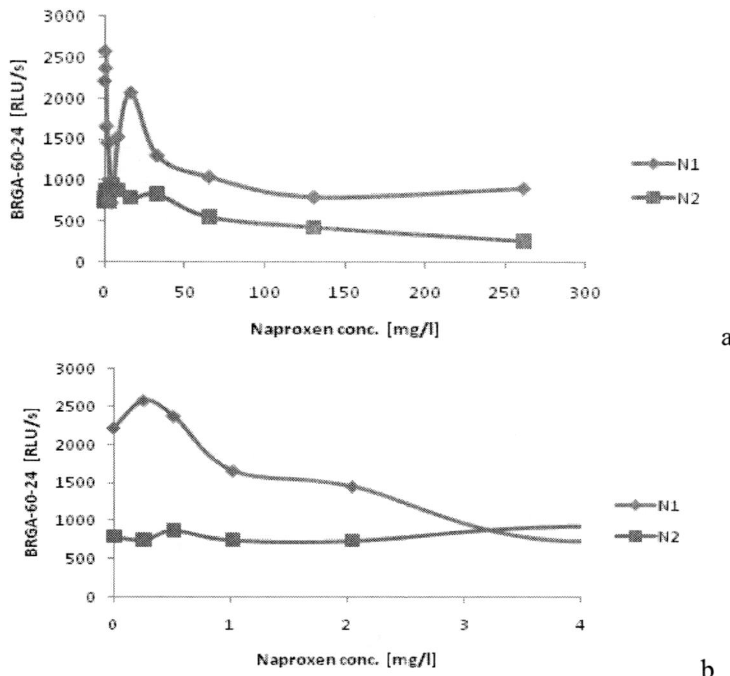

Figure 6. Inhibition of blood ROS generation by naproxen in BRGA-60-24. The BRGA-60-24 (BRGA-60- with 24 min reaction time) was performed for blood samples N1 and N2. The approx. IC50 was 3 mg/l naproxen for N1 and 130 mg/l naproxen for N2. An approx. IC75 did not appear.

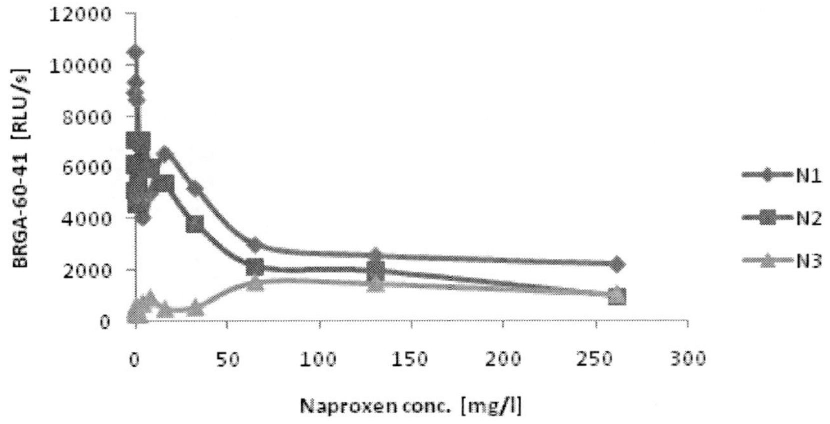

Figure 7. (Continued)

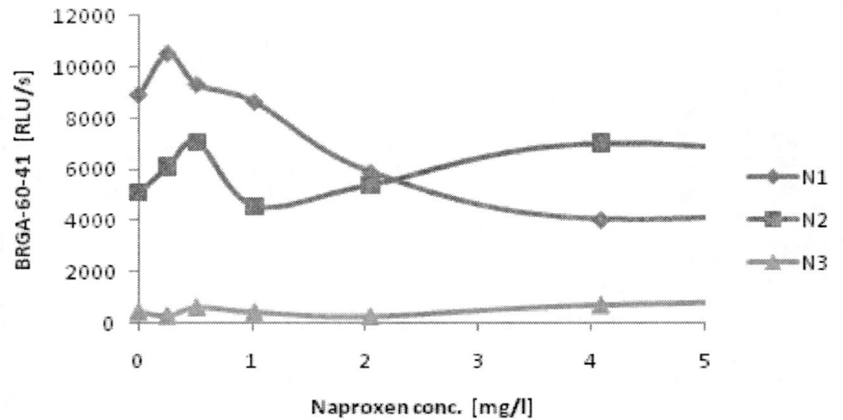

Figure 7. Modulation of blood ROS generation by naproxen in BRGA-60-41. The BRGA-60-41 (BRGA-60- with 41 min reaction time) was performed. The approx. IC50 was 4 mg/l naproxen for N1 and 50 mg/l naproxen for N2. The approx. IC75 was 200 mg/l naproxen. For N3 the approx. SC200 was 8 mg/l naproxen.

REFERENCES

[1] Burmester G, Lanas A, Biasucci L, Hermann M, Lohmander S, Olivieri I, Scarpignato C, Smolen J, Hawkey C, Bajkowski A, Berenbaum F, Breedveld F, Dieleman P, Dougados M, MacDonald T, Mola EM, Mets T, Van den Noortgate N, Stoevelaar H. The appropriate use of non-steroidal anti-inflammatory drugs in rheumatic disease: opinions of a multidisciplinary European expert panel. *Ann Rheum Dis*. 2011; 70: 818-22.

[2] McGettigan P, Henry D. Cardiovascular risk with non-steroidal anti-inflammatory drugs: systematic review of population-based controlled observational studies. *PLoS Med*. 2011; 8: e1001098.

[3] Varas-Lorenzo C, Riera-Guardia N, Calingaert B, Castellsague J, Pariente A, Scotti L, Sturkenboom M, Perez-Gutthann S. Stroke risk and NSAIDs: a systematic review of observational studies. *Pharmacoepidemiol Drug Saf.* 2011; 20: 1225-36.

[4] Tegeder I, Geisslinger G. Cardiovascular risk with cyclooxygenase inhibitors: general problem with substance specific differences? *Naunyn Schmiedebergs Arch Pharmacol*. 2006; 373:1-17.

[5] Stief T, Brödje D. Thrombin generation by naproxen. *Hemost Lab.* 2013; 6: (issue 4)

[6] Stief T. The routine blood ROS generation assay (BRGA) triggered by typical septic concentrations of zymosan A. *Hemostasis Laboratory* 2013; 6: 89-98.

[7] Stief T. Micro-thrombi stimulate blood ROS generation. *Hemostasis Laboratory* 2013; 6 (issue 2-3)

[8] Agúndez JA, García-Martín E, Martínez C. Genetically based impairment in CYP2C8- and CYP2C9-dependent NSAID metabolism as a risk factor for gastrointestinal bleeding: is a combination of pharmacogenomics and metabolomics required to improve personalized medicine? *Expert Opin Drug Metab Toxicol.* 2009; 5: 607-20. Rodrigues AD. Impact of CYP2C9 genotype on pharmacokinetics: are all cyclooxygenase inhibitors the same? *Drug Metab Dispos*. 2005; 33: 1567-75.

[9] van Leeuwen JS, Unlü B, Vermeulen NP, Vos JC. Differential involvement of mitochondrial dysfunction, cytochrome P450 activity, and active transport in the toxicity of structurally related NSAIDs. *Toxicol In Vitro*. 2012; 26: 197-205.

[10] Stief T. Acetaminophen (AAP=paracetamol) suppresses blood ROS generation at an IC50 of only 1-3 mg/l. *Hemostasis Laboratory* 2013; 6 (issue 2-3)

[11] Stief T. Diclofenac suppresses blood ROS generation. *Hemostasis Laboratory* 2013; 6 (issue 2-3)

[12] Stief TW. Neutrophil granulocytes in hemostasis. *Hemostasis Laboratory* 2008; 1: 269-89.

[13] Stief T. Singlet oxygen ($^1O_2^*$) primes blood neutrophils to generate ROS. *Hemostasis Laboratory* 2013; 6 (issue 4)

[14] Costa D, Gomes A, Lima JL, Fernandes E. Singlet oxygen scavenging activity of non-steroidal anti-inflammatory drugs. *Redox Rep.* 2008; 13: 153-60.

[15] Costa D, Moutinho L, Lima JL, Fernandes E. Antioxidant activity and inhibition of human neutrophil oxidative burst mediated by arylpropionic acid non-steroidal anti-inflammatory drugs. *Biol Pharm Bull*. 2006; 29: 1659-70.

[16] Paino IM, Ximenes VF, Fonseca LM, Kanegae MP, Khalil NM, Brunetti IL. Effect of therapeutic plasma concentrations of non-steroidal anti-inflammatory drugs on the production of reactive oxygen species by activated rat neutrophils. *Braz J Med Biol Res*. 2005; 38: 543-51.

[17] Parij N, Nagy AM, Fondu P, Nève J. Effects of non-steroidal anti-inflammatory drugs on the luminol and lucigenin amplified chemiluminescence of human neutrophils. *Eur J Pharmacol.* 1998; 352: 299-305.

[18] Stief TW. Drug - induced thrombin generation: the breakthrough. *Hemostasis Laboratory* 2010; 3: 3-6.

[19] Stief TW. Xenobiotic - induced pancreas carcinoma following mesenteric vein thrombosis : a hypothesis. *Hemost Lab.* 2010; 3: 253-8.

[20] Stief TW. Thrombin generation by paracetamol. *Hemostasis Laboratory* 2011; 4: 165-74.

Chapter 8

THERAPEUTIC HUMAN ANTITHROMBIN-3 INHIBITS BLOOD ROS GENERATION

ABSTRACT

Background: Antithrombin-3 (AT-3) is the main inactivator of thrombin, the main enzyme of hemostasis. AT-3 is of extraordinary importance in controlling excessive thrombin generation. Patients with pathologically low AT-3 activities in blood are at risk to develop pathologic disseminated intravascular coagulation (PIC). Patients in PIC, especially patients in the severe PIC forms PIC-1 (frequent) and PIC-2 (seldom) often require infusion of therapeutic human AT-3 combined with low-molecular-weight-heparin (LMWH) prophylaxis to re-establish the normal regulation of systemic thrombin generation. The action of different AT-3 drugs on secondary hemostasis has been studied previously. Here its action on tertiary hemostasis and inflammation is investigated.

Material and Methods: 0-5 IU/ml (final) therapeutic AT-3 a) Anbinex®-Grifols, b) AT III®-Baxter, c) Atenativ®-octapharma, or d) Kybernin P®-CSL Behring in 0.9% NaCl in 125 µl Hanks′ Balanced Salt Solution, 10 µl normal citrated blood, 10 µl 0.26 mM (final) luminol in black high quality polystyrene microwells (Brand®781608) were triggered in triplicate by 10 µl 1.9 µg/ml (final) zymosan A (ZyA) in the blood ROS generation assay (BRGA). The plate was incubated for 0-271 min at 37°C in a sterile incubator and repetitively measured in a photons-multiplyer microtiter plate luminometer (LUmo). The experiment was also performed in presence of 0.6 IU/ml (final) LMWH enoxaparin.

Results: In BRGA-40 (100% = 310 RLU/s) the approx. IC50 was 0.4 IU/ml Baxter-AT3, 0.6 IU/ml Grifols-AT3, 1.3 IU/ml Behring-AT3, or 1.3 IU/ml octapharma-AT3. In presence of 0.6 IU/ml LMWH in BRGA-56 (100% = 654 RLU/s) the approx. IC50 was 0.005 IU/ml Baxter-AT3 or octapharma-AT3, 0.01 IU/ml Grifols-AT3, 0.02 IU/ml Behring-AT3.

Discussion: In the clinically relevant added concentrations around 1 IU/ml all AT-3 drugs inhibited blood ROS generation. Thus, AT-3 can be considered as drug against pathologic systemic inflammation. Of special importance are the BRGA findings on AT-3 in combination with the prophylactic concentration of 0.6 IU/ml enoxaparin. Of even greater importance are previous results respective altered matrix coagulation activation. The ideal AT-3 does not trigger thrombin generation by itself (measured in RECA) and inhibits systemic inflammation to some extent (measured in BRGA). From all the data together it is concluded that Baxter-AT3 currently behaved as the best AT3 drug.

Keywords: Antithrombin-3, AT-3, reactive oxygen species, ROS, singlet oxygen, blood ROS generation assay, BRGA, neutrophils

INTRODUCTION

Normal intravascular coagulation (NIC) means a balanced systemic thrombin generation [1]: there is neither excessive systemically circulating thrombin as in pathologic intravascular coagulation (PIC) nor dramatically decreased thrombin generation as in anticoagulant over-dosage. The serine protease inhibitor (serpin) antithrombin-3 (AT-3) is the primary inactivator of thrombin [2], the key enzyme of hemostasis [3]. The serpin heparin cofactor-2 (HC-2) is the secondary inactivator of thrombin and might be measured as AT-3, if the routine AT-3 assay is not optimized. HC-2 cannot compensate a deficiency in AT-3 [4].

A severe deficiency of AT-3 with AT-3 activities below 60% of normal can occur in the severe PIC forms PIC-1 (hyper-activation; frequent) and PIC-2 (consumption; seldom) [5] with dangerous circulating aggressive micro-thrombi [6] might need urgent logical correction by infusion of therapeutic human AT-3 of high quality [7,8], preferably together with low-molecular-weight-heparin (LMWH) prophylaxis [9,10] to re-establish the normal regulation of systemic thrombin generation. The action of different AT-3 drugs on plasmatic thrombin generation (secondary hemostasis) has been studied previously [11]. Here its action on systemic inflammation/cellular fibrinolysis [12-30] is investigated by the blood ROS generation assay (BRGA) [31].

Material and Methods

40 µl 20 IU/ml (0-5 IU/ml final) therapeutic AT-3

a) Anbinex® (Grifols, Barcelona, Spain)
b) AT III® (Baxter, Unterschleißheim, Germany)
c) Atenativ® (octapharma, Langenfeld, Germany)
d) Kybernin P® (CSL Behring, Marburg, Germany)

in 0.9% NaCl in 125 µl Hanks′ Balanced Salt Solution (HBSS; modified without phenol red; SAFC Biosciences-Sigma, Deisenhofen, Germany; article nr. 55037C-1000ML), 10 µl normal venous blood anticoagulated by 10.6 mM Na_3-citrate (drawn after written informed consent into not expired polypropylene monovettes from Sarstedt, Nümbrecht, Germany; stored for 1d at 23°C), 10 µl 5 mM (0.26 mM final) luminol (Sigma, Deisenhofen, Germany; article nr. A4685-1G; 1g dissolved in 25.1 ml = 200 mM stem solution) in 0.9 % NaCl in black high quality polystyrene microwells (Brand, Wertheim, Germany; article nr. 781608) were triggered in triplicate by 10 µl 36 µg/ml zymosan A (ZyA; 1.9 µg/ml final) (Sigma; article nr. Z-4250-1G, lot nr. 27H0495) in the blood ROS generation assay (BRGA). The plate was incubated for 0-271 min at 37°C in a sterile incubator and repetitively measured in a photons-multiplyer microtiter plate luminometer (LUmo; Autobio-anthos, Krefeld, Germany) with an integration time of 0.5s per well.

The same experiment was performed after addition of 10 µl 13 IU/ml (0.6 IU/ml final) LMWH enoxaparin in 0.9% NaCl (Sanofi, Frankfurt, Germany).

The intra-assay coefficients of variation of the BRGA, triggered by 1.9 µg/ml ZyA, were less than 10%. Critical values were tested for significance using the $X^2\bar{x}$ test [32]. HBSS was 185.4 mg/l $CaCl_2 \cdot 2\, H_2O$, 200 mg/l $MgSO_4 \cdot 7\, H_2O$, 400 mg/l KCl, 60 mg/l KH_2PO_4, 350 mg/l $NaHCO_3$, 8000 mg/l NaCl, 90 mg/l Na_2HPO_4, 1000 mg/l glucose, pH 7.0-7.4. Expressed in molarity, the concentrations of the HBSS components are: 1.3 mM Ca^{2+}, 0.8 mM Mg^{2+}, 5.8 mM K^+, 143 mM Na^+, 144 mM Cl^-, 1.6 mM SO_4^{2-}, 0.4 mM $H_2PO_4^-$, 0.6 mM HPO_4^{2-}, 4.2 mM HCO_3^-, 5.6 mM glucose.

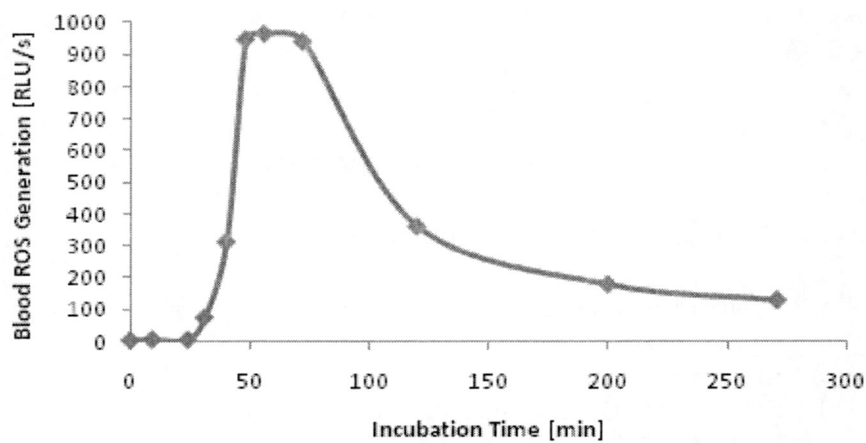

Figure 1. Kinetic of blood ROS generation in 1d old normal citrated blood. The blood ROS generation assay (BRGA) was performed with 10 µl normal citrated blood, stored for 1d at 23°C. The maximum of about 1000 RLU/s was reached at about 50 min (37°C).

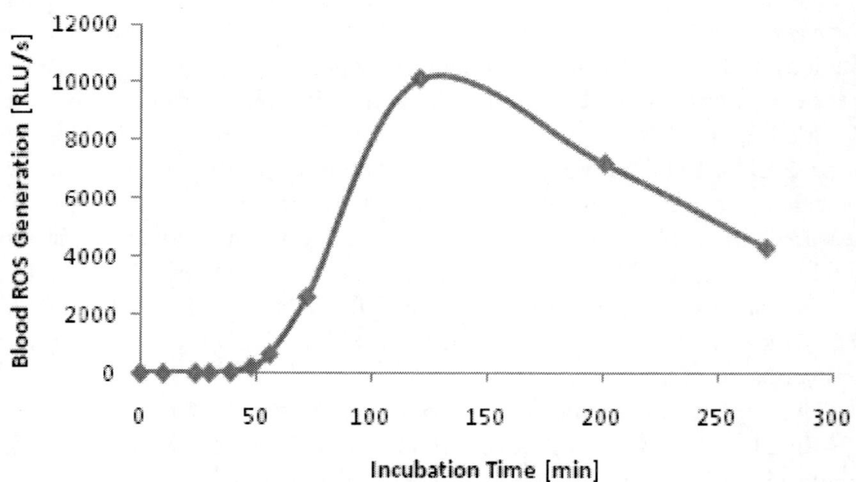

Figure 2. Kinetic of blood ROS generation in 1d old normal citrated blood in presence of 0.6 IU/ml LMWH. The blood ROS generation assay (BRGA) was performed with 10 µl normal citrated blood, stored for 1d at 23°C, in presence of 0.6 IU/ml LMWH enoxaparin. The maximum of about 11000 RLU/s was reached at about 120 min (37°C).

RESULTS

The maximum of blood ROS generation of about 1000 RLU/s was reached at about 50 min (37°C) BRGA reaction time (BRGA-50) (Figure 1). In presence of 0.6 IU/ml LMWH enoxaparin the maximum of about 11000 RLU/s [33] was reached at about 120 min (37°C) (BRGA-120) (Figure 2).

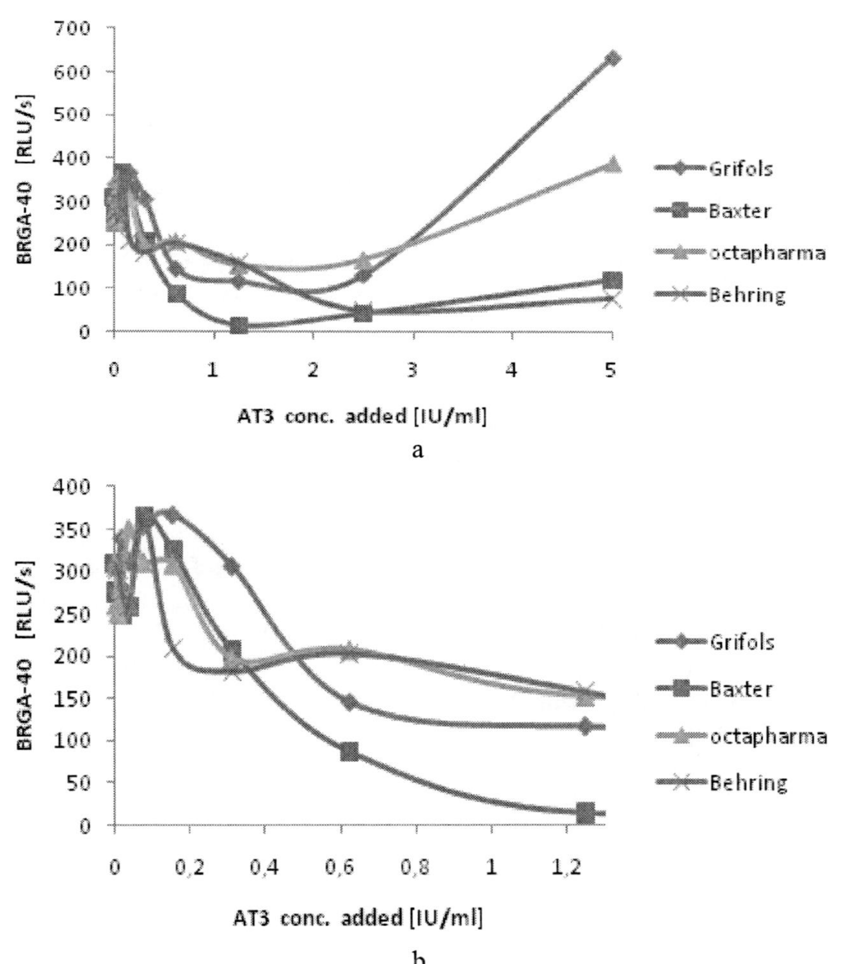

Figure 3. (Continued)

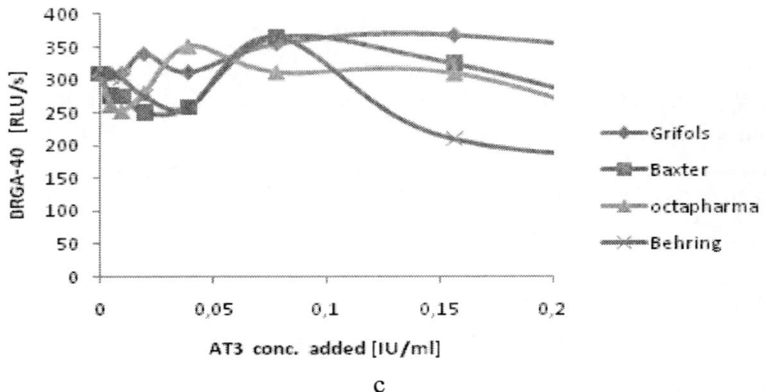

c

Figure 3. BRGA with 40 min reaction time (BRGA-40). Normal citrated blood was incubated in the BRGA in presence of 0-5 IU/ml added AT-3 of Grifols, Baxter, octapharma, or Behring.

Figure 3 demonstrates the BRGA at 40 min reaction time (BRGA-40; 100% = 310 RLU/s) in dependence of increasing concentrations of different AT-3 drugs. The approximate 50% inhibitory concentration (approx. IC50) was 0.4 IU/ml Baxter-AT3, 0.6 IU/ml Grifols-AT3, 1.3 IU/ml Behring-AT-3, or 1.3 IU/ml octapharma-AT3. In BRGA-48 (100% = 944 RLU/s) the approx. IC50 values were 0.01 IU/ml Baxter-AT3, 0.01 IU/ml Behring-AT3, 0.05 IU/ml octapharma-AT3, or 0.4 IU/ml Grifols-AT3 (Figure 4). In BRGA-56 (100% = 962 RLU/s) no AT-3 drug reached the IC50 barrier (Figure 5). Supra-therapeutic concentrations (> 2 IU/ml) of the AT-3 drugs of Grifols, Baxter, or octapharma could increase blood ROS generation (Figures 3-5). Therefore, such high AT-3 concentrations without heparin protection [34] should be avoided. In presence of 0.6 IU/ml LMWH in BRGA-56 (100% = 654 RLU/s) the approx. IC50 was 0.005 IU/ml Baxter-AT3 or octapharma-AT3, 0.01 IU/ml Grifols-AT3, 0.02 IU/ml Behring-AT3 (Figure 6). 0.63 IU/ml AT-3 added resulted in 257 RLU/s for Behring-AT3, whereas the other AT-3 drugs resulted in 122-165 RLU/s. At 0.02-0.08 IU/ml AT-3 added the Behring-AT3 resulted in about 400 RLU/s whereas the Baxter-AT3 and the octapharma-AT3 gave 100-200 RLU/s. In BRGA-72 (100% = 2608 RLU/s) no AT-3 drug reached approx. IC50, the weakest ROS generation inhibitor being the Behring-AT3 (Figure 7). In BRGA-121 or BRGA-271 (100% = 10094 RLU/s or 4309 RLU/s) the different AT-3 drugs stimulated blood ROS generation, however they did not reach approx. SC150 (Figures 8,9).

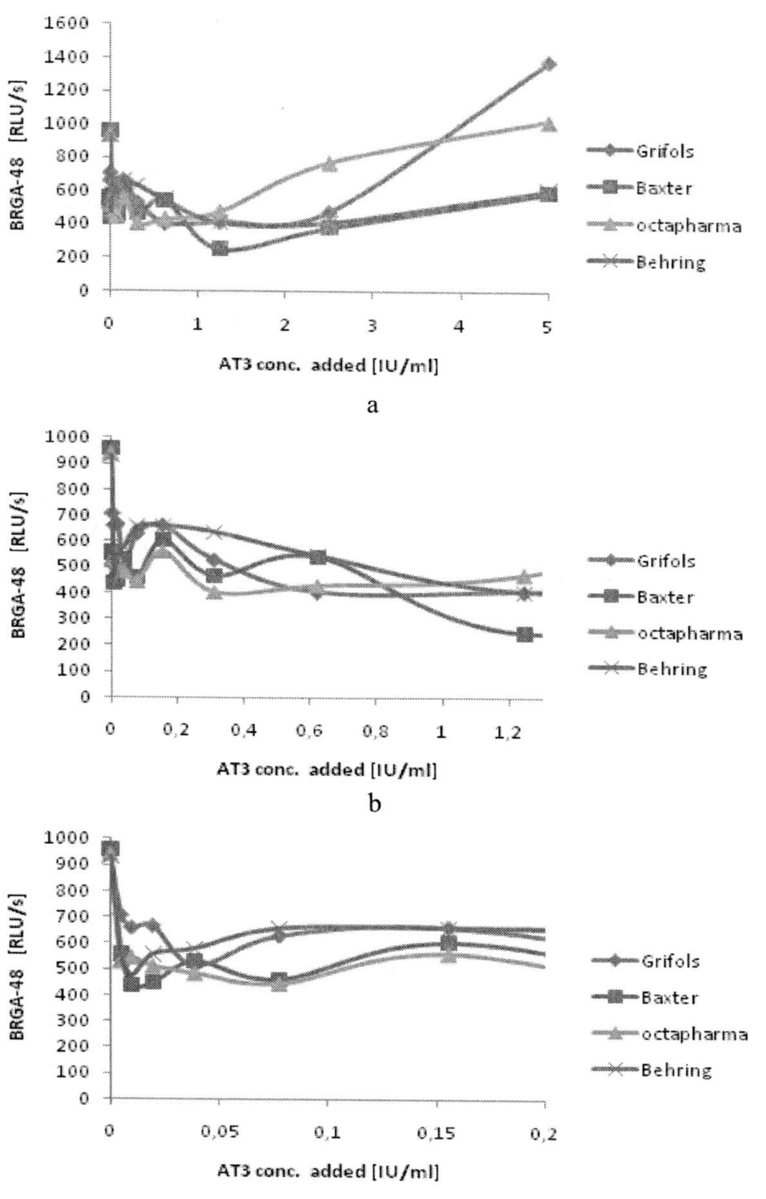

Figure 4. BRGA with 48 min reaction time (BRGA-48). Normal citrated blood was incubated in the BRGA in presence of 0-5 IU/ml added AT-3 of Grifols, Baxter, octapharma, or Behring.

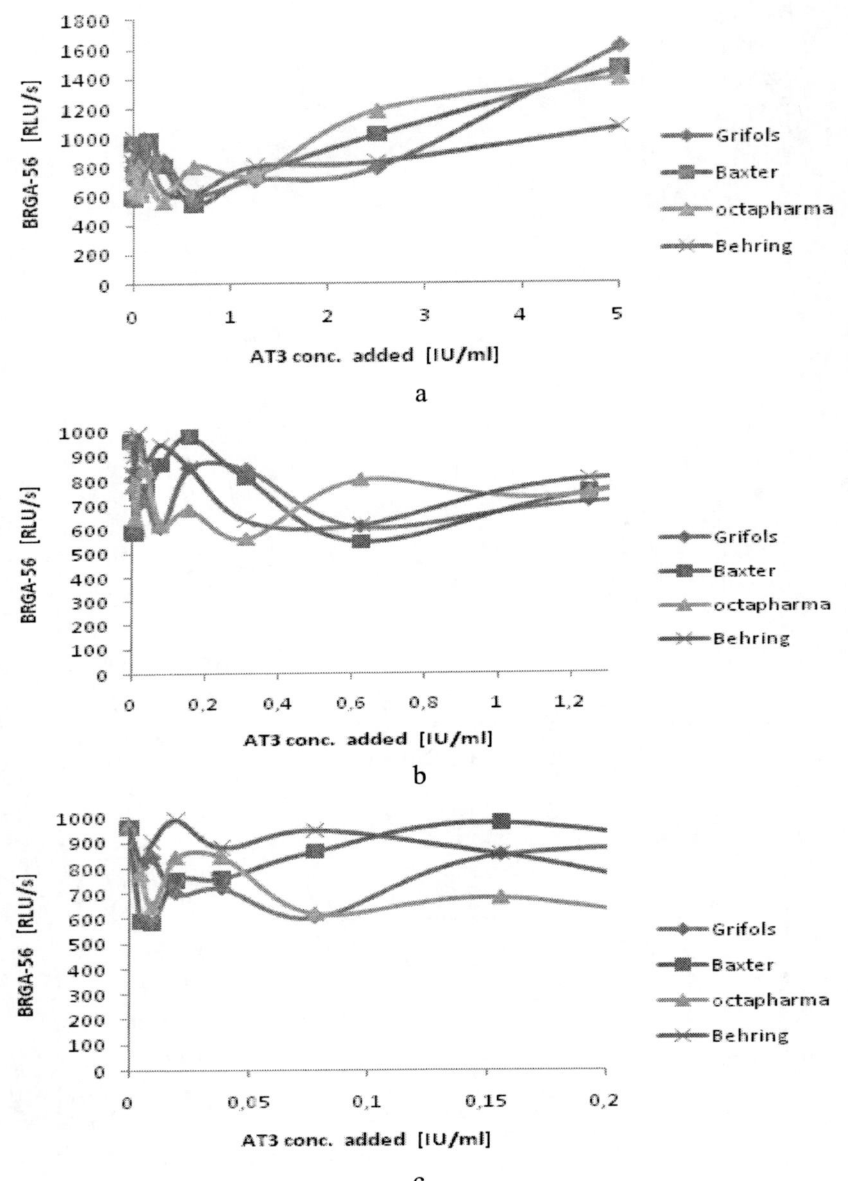

Figure 5. BRGA with 56 min reaction time (BRGA-56). Normal citrated blood was incubated in the BRGA in presence of 0-5 IU/ml added AT-3 of Grifols, Baxter, octapharma, or Behring.

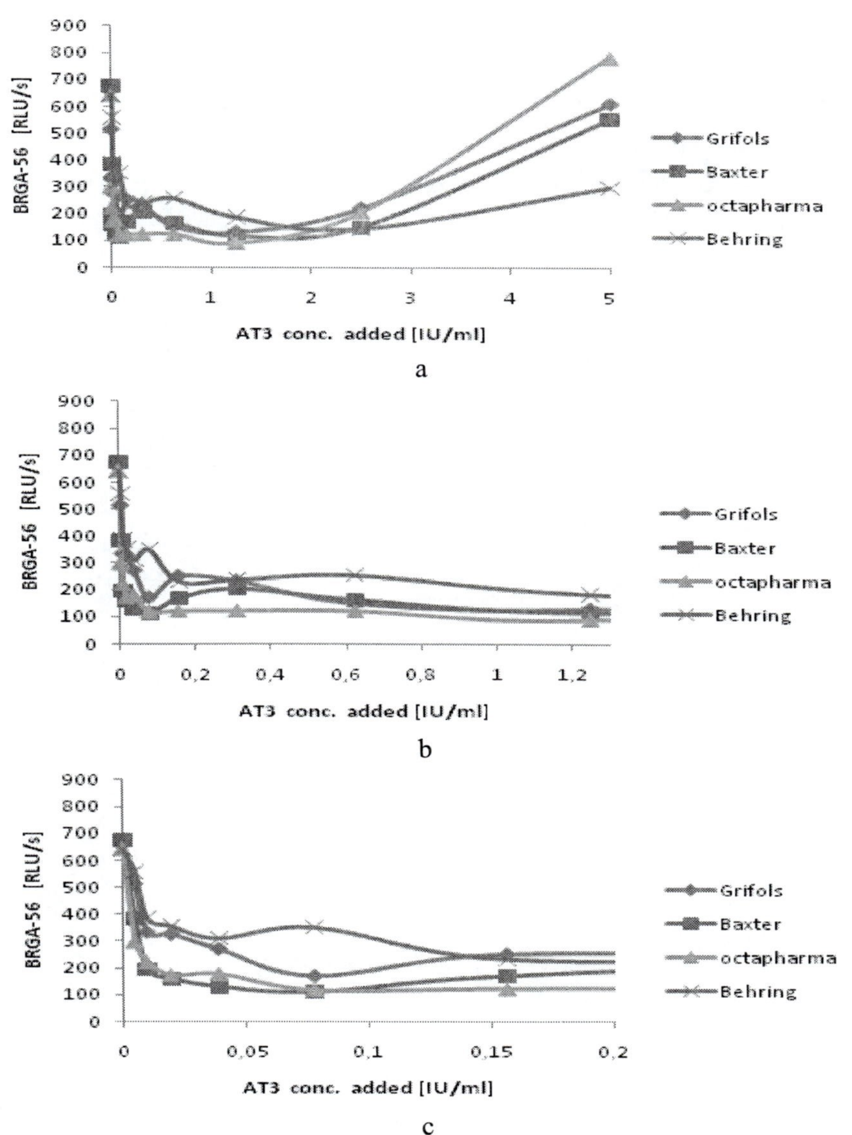

Figure 6. BRGA with 56 min reaction time (BRGA-56) in presence of 0.6 IU/ml LMWH. Normal citrated blood was incubated in the BRGA in presence of 0-5 IU/ml added AT-3 of Grifols, Baxter, octapharma, Behring and 0.6 IU/ml LMWH enoxaparin.

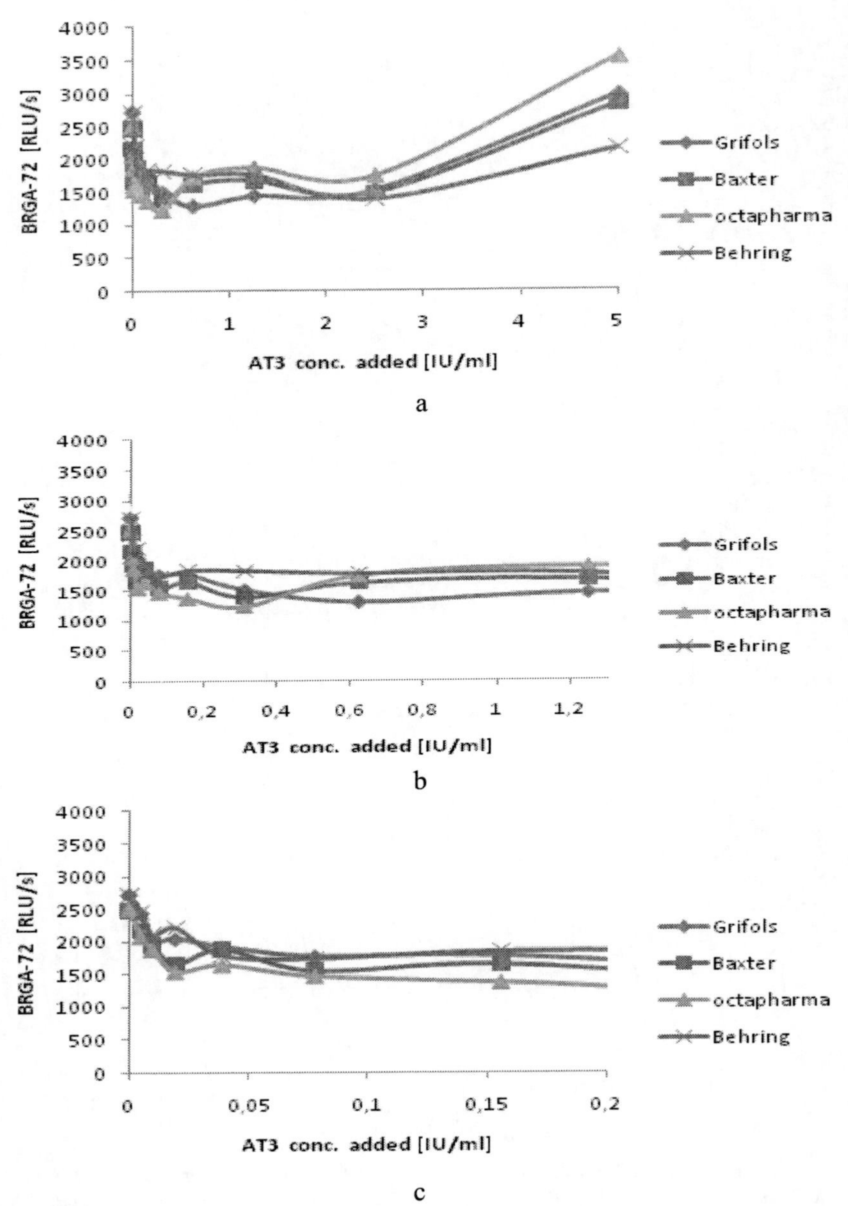

Figure 7. BRGA with 72 min reaction time (BRGA-72) in presence of 0.6 IU/ml LMWH. Normal citrated blood was incubated in the BRGA in presence of 0-5 IU/ml added AT-3 of Grifols, Baxter, octapharma, Behring and 0.6 IU/ml LMWH enoxaparin.

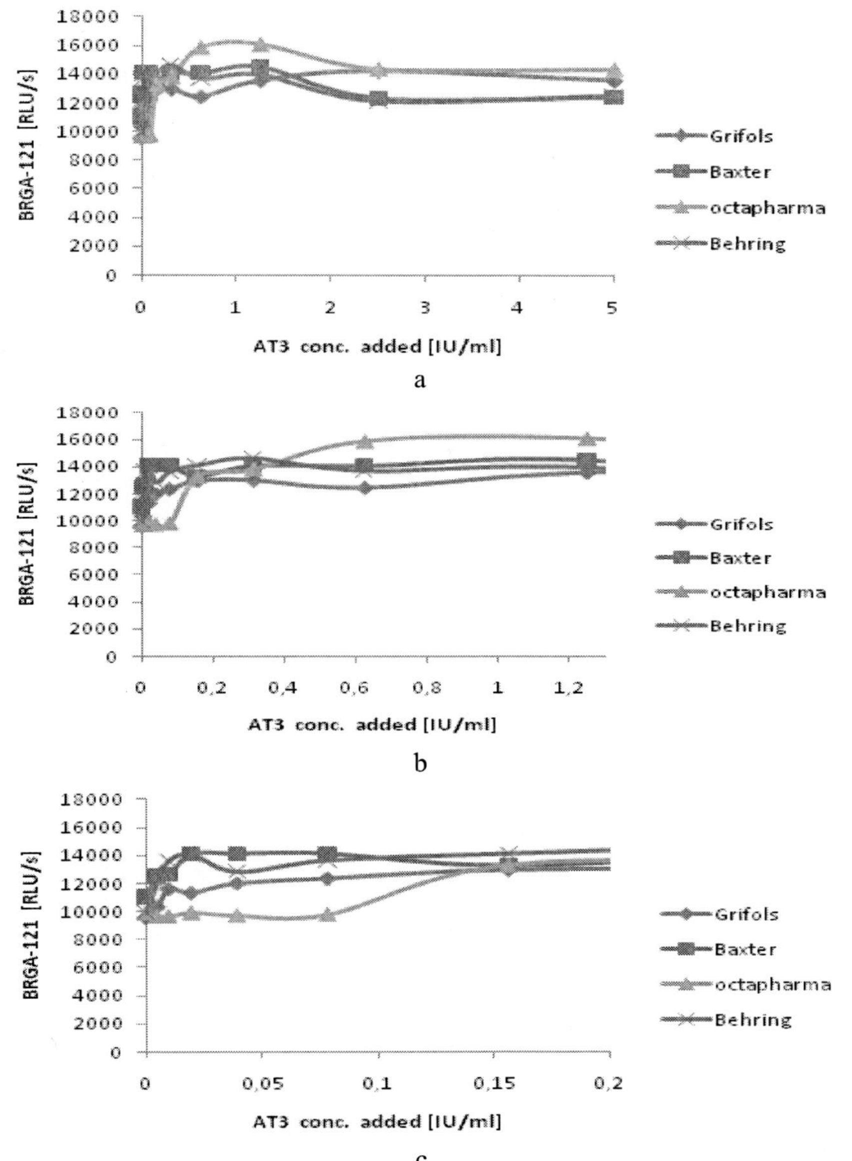

Figure 8. BRGA with 121 min reaction time (BRGA-121) in presence of 0.6 IU/ml LMWH. Normal citrated blood was incubated in the BRGA in presence of 0-5 IU/ml added AT-3 of Grifols, Baxter, octapharma, Behring and 0.6 IU/ml LMWH enoxaparin.

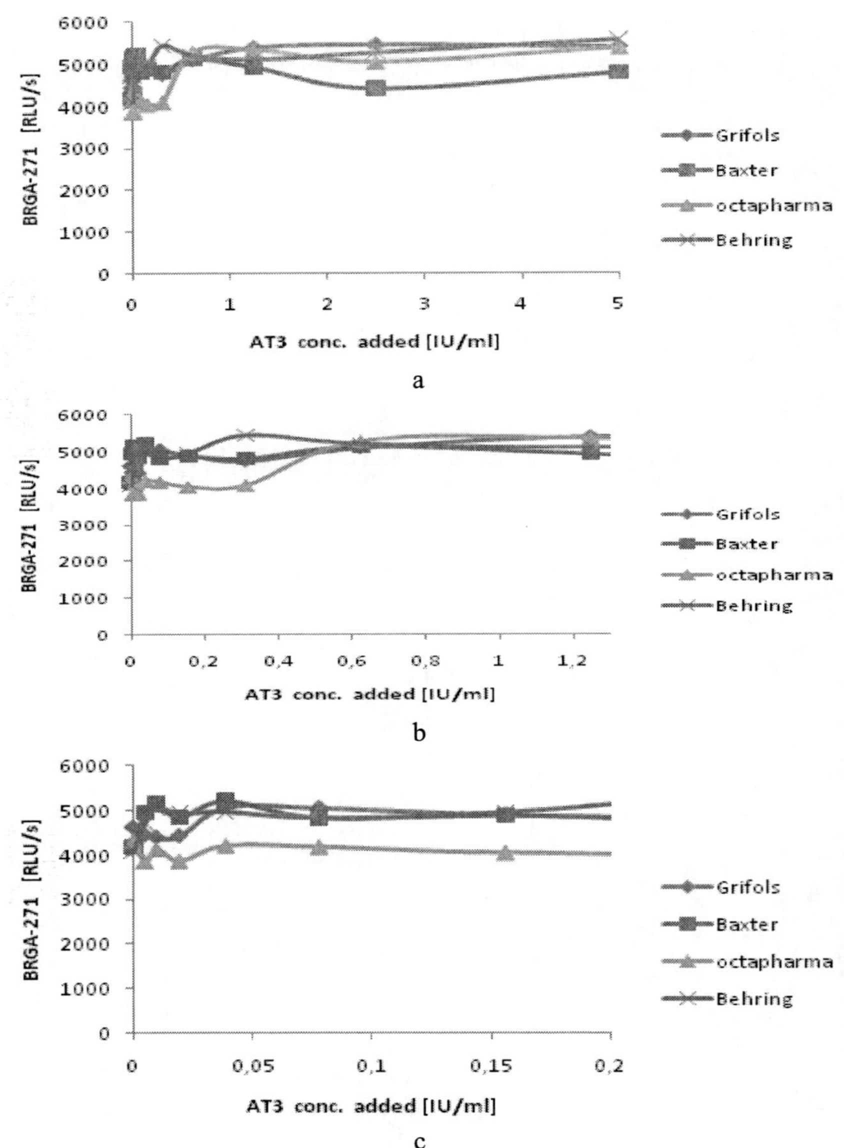

Figure 9. BRGA with 271 min reaction time (BRGA-271) in presence of 0.6 IU/ml LMWH. Normal citrated blood was incubated in the BRGA in presence of 0-5 IU/ml added AT-3 of Grifols, Baxter, octapharma, Behring and 0.6 IU/ml LMWH enoxaparin.

Discussion

The present work clearly demonstrates that there are differences between various human AT-3 drugs respective blood ROS generation modulation. In the clinically relevant added concentrations around 1 IU/ml the AT-3 drugs inhibited blood ROS generation. Thus, the AT-3 drugs have to be considered as anti-inflammatory agents [35-39]. Antithrombin-3 inhibits the increase in neutrophil stiffness upon NADPH-oxidase assembly by inhibiting actin polymerization [40].

Of special importance are the BRGA findings on AT-3 in combination with the prophylactic concentration of 0.6 IU/ml enoxaparin. Of even greater importance are the results respective altered matrix – coagulation activation [11]. The ideal AT-3 does not trigger thrombin generation by itself (measured in RECA) and inhibits systemic inflammation to some extent (measured in BRGA). From all the data together it is concluded that Baxter-AT3 currently behaved as the best one.

Acknowledgments

No specific funding, no conflict of interest.

References

[1] Stief TW. The laboratory diagnosis of the pre-phase of pathologic disseminated intravascular coagulation. *Hemostasis Laboratory* 2008; 1: 2-20.

[2] Bick RL, Bick MD, Fekete LF. Antithrombin III patterns in disseminated intravascular coagulation. *Am J Clin Pathol.* 1980; 73: 577-83.

[3] Stief TW. *Thrombin – applied clinical biochemistry of the main factor of coagulation.* In: Thrombin: function and pathophysiology. Stief T, ed.; Nova science publishers; New York; 2012; pp. vii-xx.

[4] Stief TW. Antithrombin III determination in nearly undiluted plasma. *Laboratory Medicine* 2008; 39: 46-8.

[5] Bick RL. Clinical relevance of antithrombin III. *Semin Thromb Hemost.* 1982; 8: 276-87.

[6] Stief TW. Coagulation activation by lipopolysaccharides. *Clin Appl Thrombosis/Hemostasis* 2009; 15: 209-19.

[7] Bick RL. Disseminated intravascular coagulation: a clinical/laboratory study of 48 patients. *Ann N Y Acad Sci*. 1981; 370: 843-50.

[8] Bick RL. Disseminated intravascular coagulation: a review of etiology, pathophysiology, diagnosis, and management: guidelines for care. *Clin Appl Thromb Hemost*. 2002; 8: 1-31.

[9] Fareed J, Hoppensteadt DA, Bick RL. An update on heparins at the beginning of the new millennium. *Semin Thromb Hemost*. 2000; 26 Suppl 1:5-21.

[10] Lavi S, Cantor WJ, Casanova A, Tan MK, Yan AT, Džavík V, Fitchett D, Cohen EA, Borgundvaag B, Heffernan M, Ducas J, Goodman SG. Efficacy and safety of enoxaparin compared with un-fractionated heparinin the pharmacoinvasive management of acute ST-segment elevation myocardial infarction: Insights from the TRANSFER-AMI trial. *Am Heart J*. 2012; 163: 176-81.

[11] Stief TW. Thrombin generation in presence of therapeutic antithrombin 3. *Hemostasis Laboratory* 2012; 5: 91-102.

[12] Stief TW. The physiology and pharmacology of singlet oxygen. *Med Hypoth*. 2003; 60: 567-72.

[13] Stief TW. Regulation of hemostasis by singlet-oxygen ($^1\Delta O_2$). *Curr Vasc Pharmacol* 2004; 2: 357-62.

[14] Stief TW, Fareed J. The antithrombotic factor singlet oxygen/light ($^1O_2$/hv). *Clin Appl Thrombosis/Hemostasis* 2000; 6: 22-30.

[15] Stief TW. The blood fibrinolysis / deep - sea analogy: a hypothesis on the cell signals singlet oxygen / photons as natural antithrombotics. *Thromb Res*. 2000; 99: 1-20.

[16] Stief T. Light quants of low wave length prime blood neutrophils for ROS generation. *Hemostasis Laboratory* 2013; 6: (issue 4).

[17] Stief T. Blood neutrophils see UV light: 340 nm ultraviolet A stimulates blood ROS generation nearly half as strong as 405 nm violet photons. *Hemostasis Laboratory* 2013; 6 (issue 4).

[18] Stief T. Blood neutrophils alert each other by photons. *Hemostasis Laboratory* 2013; 6 (issue 4).

[19] Stief TW. Neutrophil granulocytes in hemostasis. *Hemostasis Laboratory* 2009; 2: 269-89.

[20] Weiss SJ, Lampert MB, Test ST. Long-lived oxidants generated by human neutrophils: characterization and bioactivity. *Science*. 1983; 222: 625-8.

[21] Stief TW. Singlet oxygen potentiates thrombolysis. *Clin Appl Thrombosis/Hemostasis* 2007; 13: 259-78.

[22] Stief TW. Modulation of granulocyte mediated thrombolysis. *Hemostasis Laboratory* 2008; 1: 77-102.

[23] Stief T. Micro-thrombi stimulate blood ROS generation. *Hemostasis Laboratory* 2013; 6 (issue 2-3)

[24] Goodridge HS, Wolf AJ, Underhill DM. Beta-glucan recognition by the innate immune system. *Immunol Rev*. 2009; 230: 38-50.

[25] Nathan CF. Neutrophil activation on biological surfaces. Massive secretion of hydrogen peroxide in response to products of macrophages and lymphocytes. *J Clin Invest*. 1987; 80: 1550-60.

[26] Seguchi H, Kobayashi T. Study of NADPH oxidase-activated sites in human neutrophils. *J Electron Microsc*. 2002; 51: 87-91.

[27] Rosen H, Klebanoff SJ. Formation of singlet oxygen by the myeloperoxidase-mediated antimicrobial system. *J Biol Chem*. 1977; 252: 4803-10.

[28] Kiryu C, Makiuchi M, Miyazaki J, Fujinaga T, Kakinuma K. Physiological production of singlet molecular oxygen in the myeloperoxidase-H_2O_2-chloride system. *FEBS Lett*. 1999; 443: 154-8.

[29] Maghzal G, Krause KH, Stocker R, Jaquet V. Detection of reactive oxygen species derived from the family of NOX (NADPH oxidases). *Free Radic Biol Med*. 2012; 53: 1903-18.

[30] Briviba K, Klotz LO, Sies H. Toxic and signaling effects of photochemically or chemically generated singlet oxygen in biological systems. *Biol Chem*. 1997; 378: 1259-65.

[31] Stief T. The routine blood ROS generation assay (BRGA) triggered by typical septic concentrations of zymosan A. *Hemost Lab*. 2013; 6: 89-98.

[32] Stief TW. The fibrinogen antigenic turbidimetric assay (FIATA). The $X^2\bar{x}$ test: the corrected chi-square comparison against the control-mean. *Clin Appl Thrombosis/Hemostasis* 2007; 13: 73-100.

[33] Stief T. LMWH stimulates blood ROS generation. *Hemostasis Laboratory* 2013; 6 (issue 2-3)

[34] Fuse S, Tomita H, Yoshida M, Hori T, Igarashi C, Fujita S. High dose of intravenous antithrombin III without heparin in the treatment of disseminated intravascular coagulation and organ failure in four children. *Am J Hematol*. 1996; 53: 18-21.

[35] Bergenfeldt M, Ohlsson K. Protease-antiprotease levels and whole-blood chemiluminescence in acute peritonitis. *Gastroenterol Jpn.* 1993; 28: 687-98.

[36] Duru S, Koca U, Oztekin S, Olguner C, Kar A, Coker C, Ulukuş C, Taşcł C, Elar Z. Antithrombin III pretreatment reduces neutrophil recruitment into the lung and skeletal muscle tissues in the rat model of bilateral lower limb ischemia and reperfusion: a pilot study. *Acta Anaesthesiol Scand.* 2005; 49: 1142-8.

[37] Okajima K. Antithrombin prevents endotoxin-induced pulmonary vascular injury by inhibiting leukocyte activation. *Blood Coagul Fibrinolysis.* 1998; 9 Suppl 2: S25-37.

[38] Opal SM, Kessler CM, Roemisch J, Knaub S. Antithrombin, heparin, and heparan sulfate. *Crit Care Med.* 2002; 30(5 Suppl): S325-31.

[39] Dunzendorfer S, Kaneider N, Rabensteiner A, Meierhofer C, Reinisch C, Römisch J, Wiedermann CJ. Cell-surface heparan sulfate proteoglycan-mediated regulation of human neutrophil migration by the serpin antithrombin III. *Blood.* 2001; 97: 1079-85.

[40] Saito H, Minamiya Y, Kalina U, Saito S, Ogawa J. Effect of antithrombin III on neutrophil deformability. *J Leukoc Biol.* 2005; 78: 777-84.

Chapter 9

THROMBIN GENERATION BY NAPROXEN

Thomas Stief and Dörte Brödje

ABSTRACT

Background: Naproxen is a non-steroidal anti-inflammatory drug (NSAID). Compared with other NSAIDs naproxen is associated with less myocardial infarctions or apoplexias. NSAIDs could enhance the circulating amount of micro-thrombi by activation of AM-coagulation (AM = altered matrix = contact phase of secondary hemostasis) generating F12a, kallikrein, and thrombin or by inhibition of cellular fibrinolysis with its central component singlet oxygen. The recalcified coagulation activity assay (RECA) is the best plasma test for spurious activations of AM-coagulation.

Material and Methods: 50 µl individual normal citrated plasmas in high quality polystyrene microwells (Brand®781600) were supplemented with 0-50 mg/l naproxen (n=50) or ibuprofen (n=12) by immediate 1+1 dilution on the plates. Immediately thereafter, the RECA was performed: 5 µl 250 mM $CaCl_2$, coagulation reaction time (CRT) of 0-20 min (37°C); 100 µl 2.5 M arginine, 0.16% Triton X 100®, pH 8.6; 3 min; 25 µl 1 mM HD-CHG-Ala-Arg-pNA in 1.25 M arginine, pH 8.7; $\Delta A/t$ at 405 nm. The approximate 200% stimulatory concentrations (approx. SC200) were determined at the most sensible CRT.

Results: The approx. SC200 of naproxen on AM-coagulation was 20±27 mg/l (MV±1SD; N=50) naproxen. 24% of the normal plasmas were very susceptible to naproxen induced AM-coagulation with an approx. SC200 < 2 mg/l naproxen (0.4 mg/l = lowest approx. SC200),

30% of the plasmas had an intermediate sensibility (approx. SC200 = 2-8 mg/l naproxen), and 46% of the plasmas were rather resistant against naproxen induced AM-coagulation with an approx. SC200 ≥ 8 mg/l naproxen. For a subgroup (n=12) the approx. SC200 was 21±29 mg/l (MV±1SD) naproxen and 1.1±1.4 mg/l ibuprofen, with a good correlation between the chemically related drugs (r= 0.897).

Discussion: Typical for contact activation, there is great inter-individual variation of the susceptibility towards the respective AM-trigger. Patients with a low naproxen SC200 for recalcified thrombin generation with additional thrombosis risk factors such as insufficient blood flow or hepatic insufficiency need low-molecular-weight-heparin (LMWH) prophylaxis.

Keywords: Naproxen, NSAID, F12a, kallikrein, thrombin, RECA

INTRODUCTION

The non-steroidal anti-inflammatory drug (NSAID) naproxen (Figure 1), eventually combined with a proton pump inhibitor [1-6] is of special interst in chronic pain treatment or in acute pain treatment of thrombophilic patients [7-9], i.e. patients with blood stasis, altered blood composition, or altered vessel walls (Virchow's Law). The NSAIDs inhibit the synthesis of prostaglandins (Figures 2-6). Prostaglandins are cell hormones that are derived of 20 carbon-atoms (eicosanoids) polyunsaturated fatty acids with 3, 4, or 5 C=C alkene groups (DGLA, AA, or EPA, respectively). Usually, phospholipase A_2 releases AA out of (membranous) phospholipids (or out of diacylglycerol = DAG). Cyclooxygenases (COX) and peroxidase oxidize AA to prostaglandin H_2 (PGH_2) (Figure 5). COX-1 generates a baseline level of prostaglandins, COX-2 is the inducible enzyme form, e.g. by inflammatory stimuli. Prostaglandin H_2 is the mother substance that enters either the cyclooxygenase pathway or the lipoxygenase pathway to form either prostaglandin/thromboxane (PGD, PGE, PGF, TXA) (Figure 6) or leukotriene, respectively. The lipoxygenase enzyme pathway is especially active in granulocytes and in macrophages [10].

Unselective NSAIDs such as ibuprofen or diclofenac inhibit both the basal level cyclooxygenase (COX-1) and the inducible COX-2. These unselective NSAISs or (here even worse) the selective COX-2 inhibitors are associated with much higher frequencies of serious thrombotic events than the unselective NSAID naproxen. Naproxen had only 50% of the risk of stroke as compared with ibuprofen [15].

Cyclooxygenase (COX) seems to be very important in cellular fibrinolysis, where the neutrophils are strongly activated preferably also via COX-2 [16-19]. NSAIDs could enhance the circulating amount of micro-thrombi by activation of AM-coagulation (AM = altered matrix = contact phase of secondary hemostasis) or by inhibition of cellular fibrinolysis (tertiary hemostasis), i.e. increasing systemic thrombin or decreasing systemic plasmin, respectively. The recalcified coagulation activity assay (RECA) is the best plasma test for spurious activations of AM-coagulation [20,21]. The blood ROS generation assay (BRGA) is the best blood test for the generation of the central compound of cellular fibrinolysis, that is singlet oxygen [22]. Here naproxen's stimulation of AM-coagulation was investigated in RECA. There is only very limited information respective the modulation of human hemostasis by naproxen, an inhibitor of pro-aggregatory TXA_2 [23].

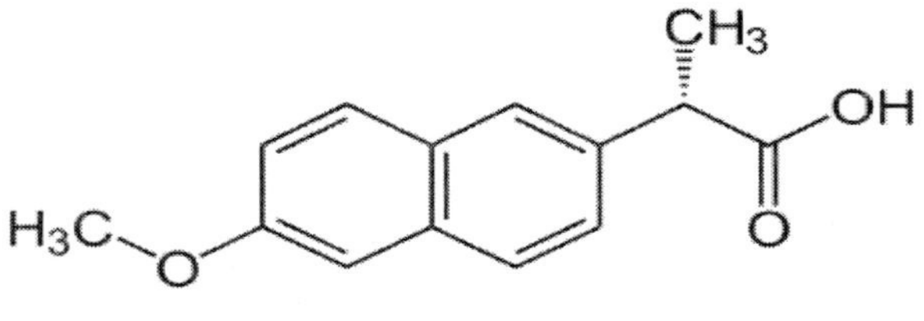

CH_3 OH H_3C O O

Figure 1. Chemical structure of naproxen [10]. 2-(6-methoxynaphthalene-2-yl) propanoic acid. MW: 230.26 Daltons; in the United Kingdom, 250 mg or 500 mg tablets are usual, with a maximum daily dose about 1000 mg. A typical plasma concentration of 160 μM corresponds to 37 mg/l. The bioavailability after oral ingestion is 95%, 99% is protein bound, 99% is hepatically metabolized to 6-desmethylnaproxen. Naproxen is a poor substrate of the cytochrome P450 system [11-13] and relatively harmless to hepatocytes [14]. Its half-life is 12-24h, eliminated by the kidneys [10].

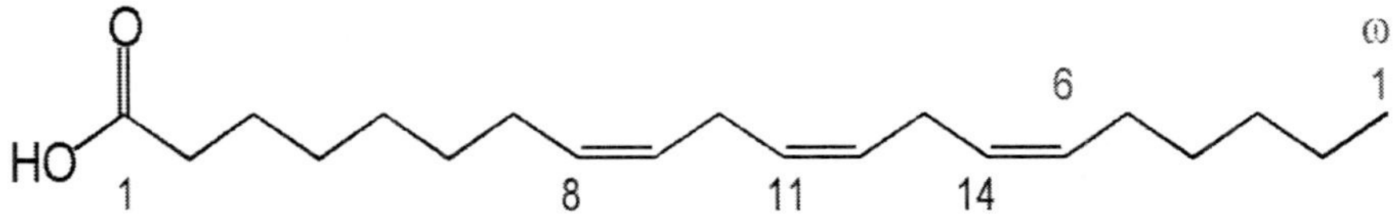

O ω 6 1 HO 1 8 11 14

Figure 2. Chemical structure of dihomo-γ-linolenic acid (DGLA = 20:3) [10]. *cis,cis,cis*-8,11,14-eicosatrienoic acid (DGLA); 20:3 (ω−6); $C_{20}H_{34}O_2$.

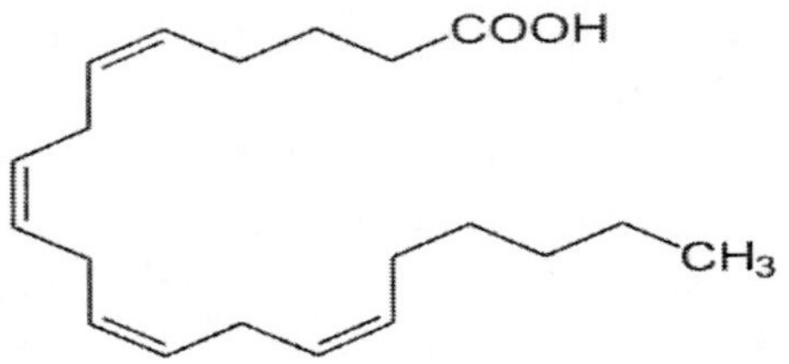

Figure 3. Chemical structure of arachidonic acid (AA = 20:4) [10]. All cis 5,8,11,14-eicosatrienoic acid (DGLA); 20:4 (ω−6); $C_{20}H_{32}O_2$. Action of delta-5-desaturase on DGLA generates AA. COX-1/COX-2, lipoxygenases, or CYP450 dependent epoxygenases generate prostaglandins, leukotrienes, or epoxy-eicosa-trienic acids.

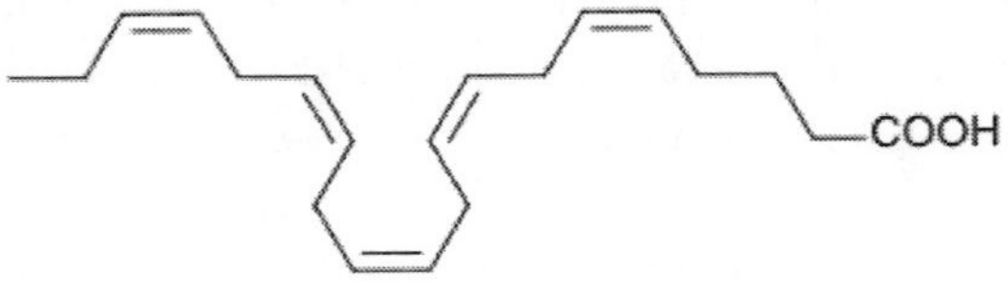

Figure 4. Chemical structure of eicosapentaenoic acid (EPA =20:5) [10]. All cis-5,8,11,14,17-eicosatrienoic acid (EPA); 20:5 (ω−3); $C_{20}H_{30}O_2$.

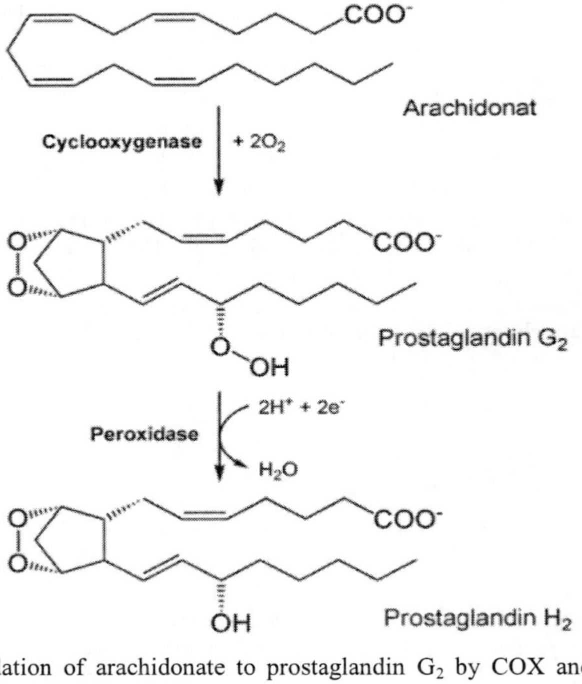

Figure 5. Oxidation of arachidonate to prostaglandin G_2 by COX and reduction of PGG_2 to PGH_2 by peroxidase [10].

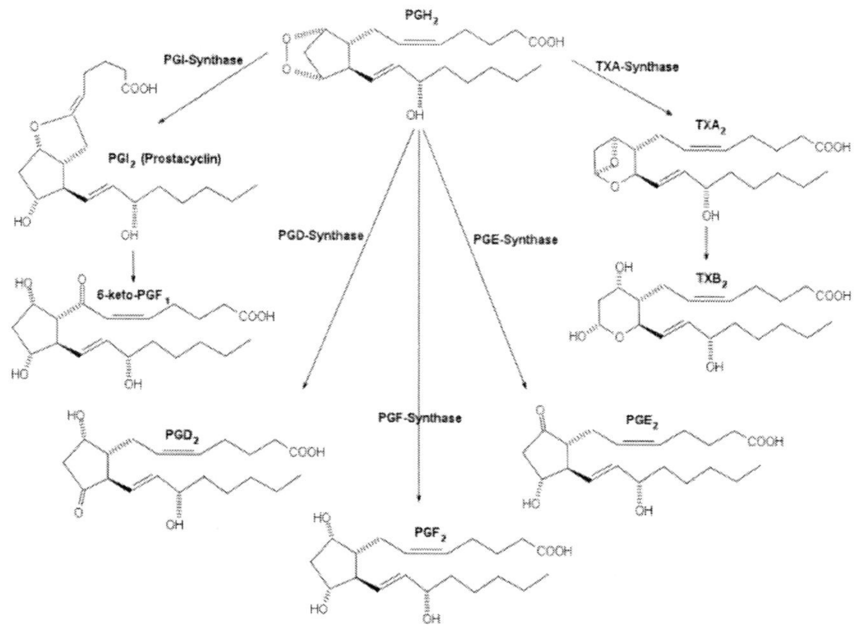

Figure 6. Generation of prostaglandins/thromboxane out of PGH_2 [10]. The different prostaglandins PGE_2, PGF_2, PGD_2, PGI_2 (prostacyclin with its endothelial metabolite 6-keto-PGF_1) and thromboxane (TXA_2, TXB_2) are generated by specific enzymes out of the mother prostaglandin PGH_2.

Material and Methods

50 µl samples of individual normal -30°C frozen/23°C thawed platelet poor citrated plasma from 50 healthy donors (4.5 ml venous blood drawn into polypropylene monovettes pre-filled with 0.5 ml 106 mM Na_3-citrate, pH 7.4; Sarstedt, Nümbrecht, Germany; centrifuged within 2h at 2800g for 10 min at 23°C) were given after written informed consent into high quality polystyrene U-wells (Brand, Wertheim, Germany; article nr. 781600). The plasma samples were immediately supplemented with 0-50 mg/l (final) naproxen (Sigma-Fluka, Deisenhofen, Germany) by repetitive 1+1 dilution on the microtiter plates.

Immediately thereafter, the recalcified coagulation activity assay (RECA) was performed in duplicate: 5 µl 250 mM $CaCl_2$ were added, the thrombin generation was stopped after coagulation reaction times (CRT) of 0, 10, or 20 min (37°C) by addition of 100 µl 2.5 M arginine, 0.16% Triton X 100®, pH 8.6 (Sigma). After 3 min 25 µl 1 mM chromogenic thrombin substrate HD-CHG-Ala-Arg-pNA (Pentapharm, Basel, Switzerland) in 1.25 M arginine, pH 8.7 were added and the increase of 405 nm absorbance with time at 37°C (ΔA/t) was determined by a microtiter plate photometer with a 1 mA resolution (PHOmo; anthos, Krefeld, Germany). The maximal absorbance increase was about 1000 mA, the approximate linearity of thrombin activity and substrate cleavage is guaranteed (nearly no substrate exhaustion) up to about 40% of the maximum = 400 mA. The ΔA values were standardized by 50 µl 0.04 IU/ml bovine thrombin standards in 5% human albumin, resulting in a ΔA/h of about 100 mA (37°C). The basal thrombin activity in plasma (0 min reaction reaction with Ca^{2+} reagent = RECA-0; about 0.01 IU/ml thrombin) was individually subtracted from the measured respective thrombin activity. Considered and shown are only thrombin generations in the ascending part of the thrombin generation curve, that is in absence of significant amounts of antithrombin-1 (= nascent fibrin) [24-26]. The time interval of the coagulation reaction time (CRT) varies from 5 min to 40 min (37°C). Usually, in fresh normal plasma 20 min is the ideal CRT. Slightly pre-activated plasma needs a shorter CRT, such as 15 min; strongly or excessively pre-activated plasma need an even shorter CRT, such as 10 min or 5 min. Plasma with additives such as aprotinin (in lyophilized plasma) need a longer CRT, such as 30 min or 40 min, depending on the concentration of aprotinin added. The mean values were calculated, the intra-assay coefficients of variation were less than 10%. The approximate 200% stimulatory concentrations (approx. SC200) were determined at the most sensible CRT. If the resulting approx. SC200 was higher than 50 mg/l the plasma was assayed again with 100 mg/l naproxen as maximal trigger concentration. For control, n=12 plasmas were supplemented with 0-50 mg/l to naproxen chemically related ibuprofen – sodium salt (Sigma) or totally different AM-triggers were used.

50 µl unfrozen normal human citrated plasma (9 parts of venous blood + 1 part of 106 mM sodium citrate, within 1-2h centrifuged for 10 min (23°C) at 2800g, stored up to about 3h 23°C) of 6 healthy donors in high quality polystyrene U-wells microtiter plates were supplemented after written informed consent with 2 µl 0-5.2 mM 3-hydroxy-butyrate (3-HB sodium salt; Sigma-Aldrich, Deisenhofen, Germany), acetoacetate (lithium salt; MW: 108.02 Daltons; Sigma; article nr. A8509-250MG), or acetone (MW: 58.08 Daltons; density: 0.79; 164 mM = 9.525 mg/ml = 12.06 µl 100% acetone/ml final solution) in 0.9 % NaCl of different concentrations, resulting in a final plasma concentration of up to 200 µM ketone bodies. Immediately thereafter, the RECA was performed with 0, 10, or 20 min coagulation reaction time.

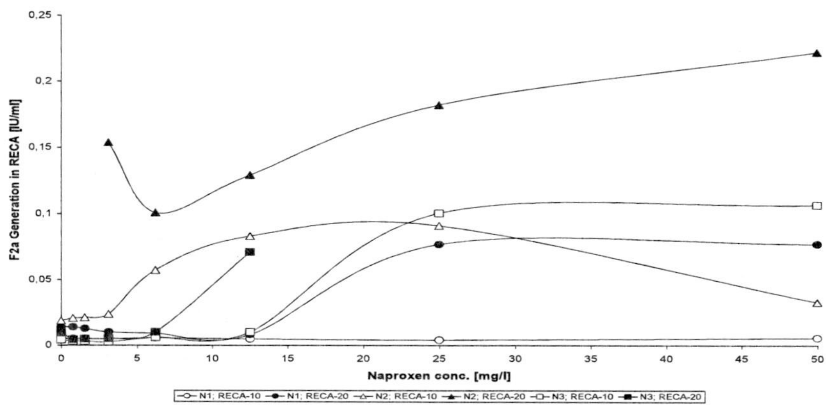

Figure 7. Thrombin (F2a) generation by naproxen in individual plasmas 1-3. 0-50 mg/l naproxen were added to 50 µl individual platelet poor citrated plasma (frozen/thawed) in high quality polystyrene U-wells (Brand®781600). The RECA was performed with 10 min (white symbols) or 20 min (black symbols) CRT (RECA-10 or RECA-20). The approx. SC200 values were 16, 4, 8 mg/l naproxen for plasmas 1 (circles), 2 (triangels), 3 (squares), respectively.

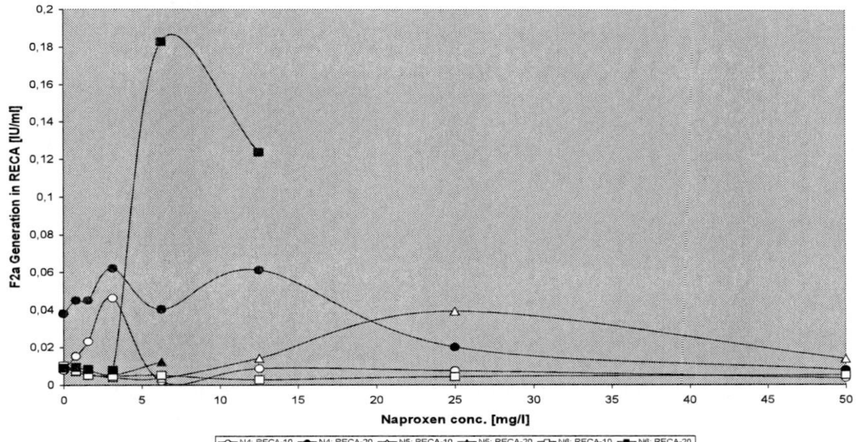

Figure 8. Thrombin (F2a) generation by naproxen in individual plasmas 4-6. 0-50 mg/l naproxen were added to 50 µl individual platelet poor citrated plasma (frozen/thawed) in high quality polystyrene U-wells (Brand®781600). The RECA was performed with 10 min (white symbols) or 20 min (black symbols) CRT (RECA-10 or RECA-20). The approx. SC200 values were 1, 15, 4 mg/l naproxen for plasmas 4 (circles), 5 (triangels), 6 (squares), respectively.

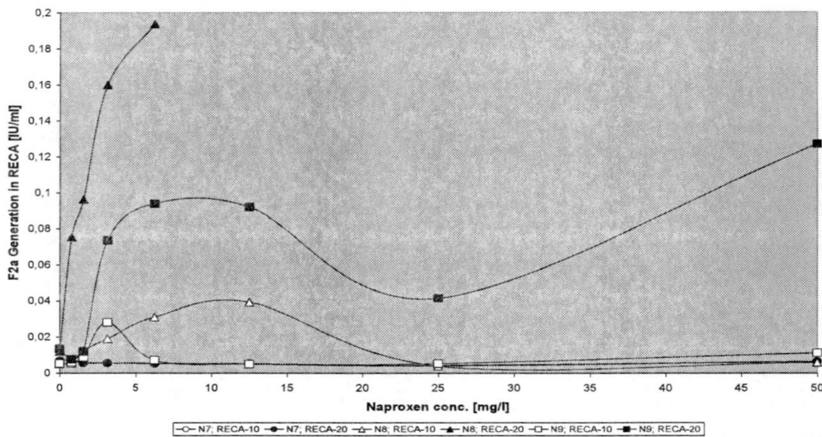

Figure 9. Thrombin (F2a) generation by naproxen in individual plasmas 7-9. 0-50 mg/l naproxen were added to 50 µl individual platelet poor citrated plasma (frozen/thawed) in high quality polystyrene U-wells (Brand®781600). The recalcified coagulation activity assay (RECA) was performed with 10 min (white symbols) or 20 min (black symbols) coagulation reaction time (RECA-10 or RECA-20). The approx. SC200 values were > 50, 0.4, 2 mg/l naproxen for plasmas 7 (circles), 8 (triangels), 9 (squares).

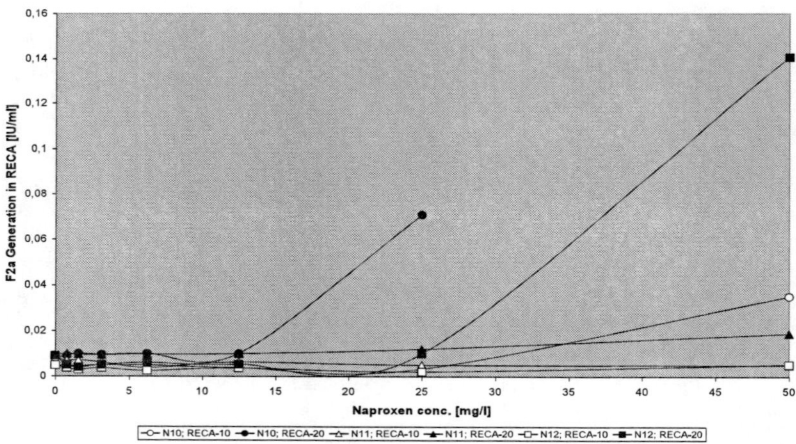

Figure 10. Thrombin (F2a) generation by naproxen in individual plasmas 10-12. 0-50 mg/l naproxen were added to 50 µl individual platelet poor citrated plasma (frozen/thawed) in high quality polystyrene U-wells (Brand®781600). The recalcified coagulation activity assay (RECA) was performed with 10 min (white symbols) or 20 min (black symbols) coagulation reaction time (RECA-10 or RECA-20). The approx. SC200 values were 15, 50, 28 mg/l naproxen for plasmas 10 (circles), 11 (triangels), 12 (squares), respectively.

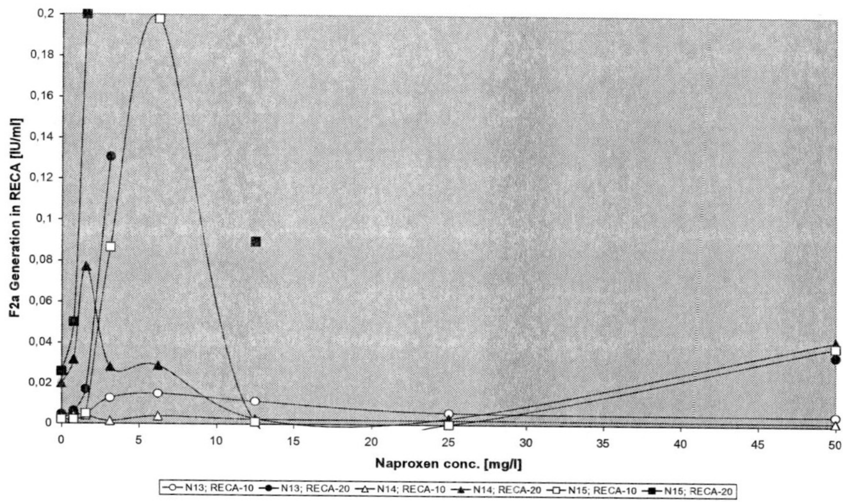

Figure 11. Thrombin (F2a) generation by naproxen in individual plasmas 13-15. 0-50 mg/l naproxen (sodium salt) were added to 50 µl individual platelet poor citrated plasma (frozen/thawed) in high quality polystyrene U-wells (Brand®781600). The recalcified coagulation activity assay (RECA) was performed with 10 min (white symbols) or 20 min (black symbols) coagulation reaction time (RECA-10 or RECA-20). The approx. SC200 values were 1 mg/l naproxen.

Results

Figures 7-10 demonstrate the determination of the approx. SC200 values of naproxen on AM-coagulation. Since the values were considerably higher than the mean SC200 found for ibuprofen (about 3 mg/l) [27,28], 12 normal invididual plasmas were supplemented either with naproxen or with ibuprofen, and the approx. SC200 was determined (Figures 11-18). The approx. SC200 values were 21±29 mg/l (MV±1SD) naproxen vs. 1.1±1.4 mg/l ibuprofen, they correlated with r= 0.897. The x/y transformation formula was nearly y=0.05x. The ibuprofen SC200 can be calculated from the naproxen SC200 dividing it by approximately 20. With such a good correlation it seemed not necessary to continue with naproxen/ibuprofen SC200 comparisons.

Figures 20-28demonstrate further results on individual normal plasmas supplemented with naproxen. Figure 29 summarizes the naproxen SC200 results: 12/50 (24%) of the plasmas were very susceptible to naproxen induced AM-coagulation with an approx. SC200 < 2 mg/l naproxen, 15/50 (30%) of the plasmas had an intermediate sensibility with an approx. SC200 of 2-8 mg/l naproxen, and 23/50 (46%) of the plasmas were rather resistant against naproxen induced AM-coagulation with an approx. SC200 ≥ 8 mg/l naproxen. In figures 30-35 ketone bodies as another AM-coagulation trigger were used as control. Normally, the approx. SC200 values are lower if the trigger is added as 2 µl volume to the plasma than by repetitive 1+1 dilution of plasma [29,30]. A typical result is that of figure 30, where the approx. SC200 values were 0.1 mM for 3-HB, 0.05 mM for acetoacetate, or 0.01 mM for acetone [31].

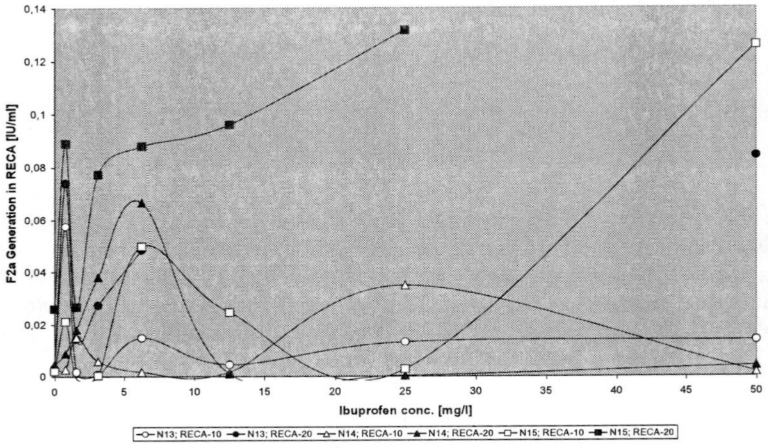

Figure 12. Thrombin (F2a) generation by ibuprofen in individual plasmas 13-15. 0-50 mg/l ibuprofen (sodium salt) were added to 50 µl individual platelet poor citrated plasma (frozen/thawed) in high quality polystyrene U-wells (Brand®781600). The recalcified coagulation activity assay (RECA) was performed with 10 min (white symbols) or 20 min (black symbols) coagulation reaction time (RECA-10 or RECA-20). The approx. SC200 values were 0.1, 0.8, 0.4 mg/l ibuprofen.

DISCUSSION

Naproxen inhibits the generation of prostaglandins not only by inhibition of COX-1 or COX-2 but also by inhibition of the hormone sensitive lipase [32]. The present work demonstrates that naproxen is indeed a relatively mild trigger of AM-coagulation. Compared with dicofenac [27] or ibuprofen [28] it triggers AM-coagulation about 50-fold or 20-fold less efficiently. This means that patients that require NSAIDs, especially patients on chronic therapy, would benefit from a change towards naproxen.

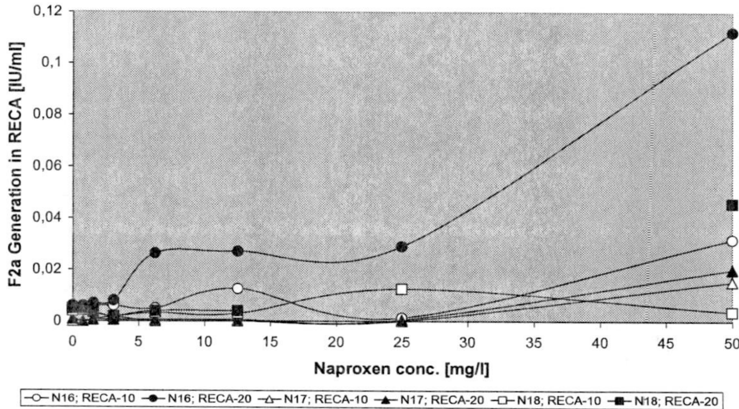

Figure 13. Thrombin (F2a) generation by naproxen in individual plasmas 16-18. 0-50 mg/l naproxen (sodium salt) were added to 50 μl individual platelet poor citrated plasma (frozen/thawed) in high quality polystyrene U-wells (Brand®781600). The recalcified coagulation activity assay (RECA) was performed with 10 min (white symbols) or 20 min (black symbols) coagulation reaction time (RECA-10 or RECA-20). The approx. SC200 values were 4, 30, 15 mg/l naproxen.

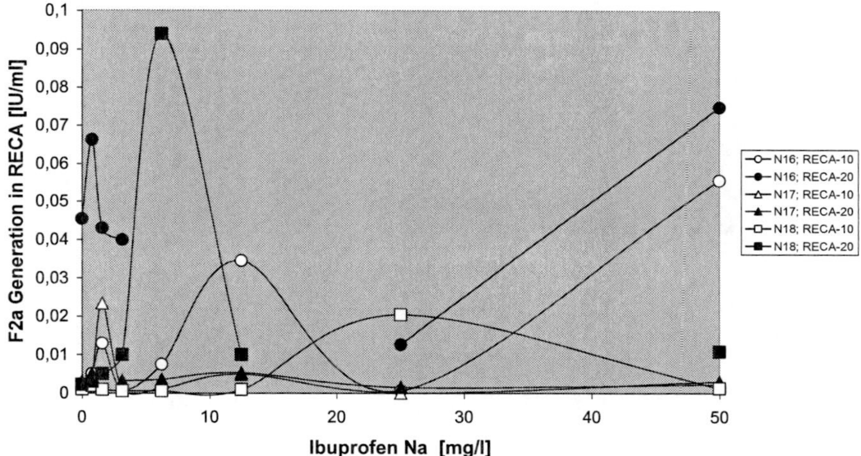

Figure 14. Thrombin (F2a) generation by ibuprofen in individual plasmas 16-18. 0-50 mg/l naproxen (sodium salt) were added to 50 μl individual platelet poor citrated plasma (frozen/thawed) in high quality polystyrene U-wells (Brand®781600). The recalcified coagulation activity assay (RECA) was performed with 10 min (white symbols) or 20 min (black symbols) coagulation reaction time (RECA-10 or RECA-20). The approx. SC200 values were 0.5, 1, 1 mg/l ibuprofen.

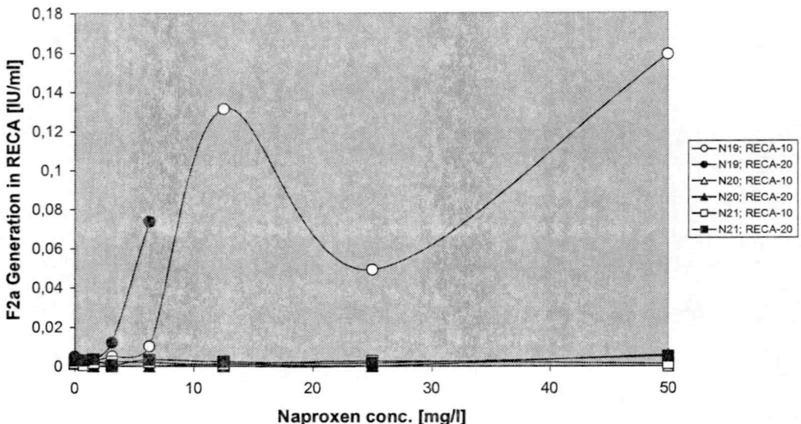

Figure 15. Thrombin (F2a) generation by naproxen in individual plasmas 19-21. 0-50 mg/l naproxen (sodium salt) were added to 50 µl individual platelet poor citrated plasma (frozen/thawed) in high quality polystyrene U-wells (Brand®781600). The recalcified coagulation activity assay (RECA) was performed with 10 min (white symbols) or 20 min (black symbols) coagulation reaction time (RECA-10 or RECA-20). The approx. SC200 values were 3, 50, 70 mg/l naproxen.

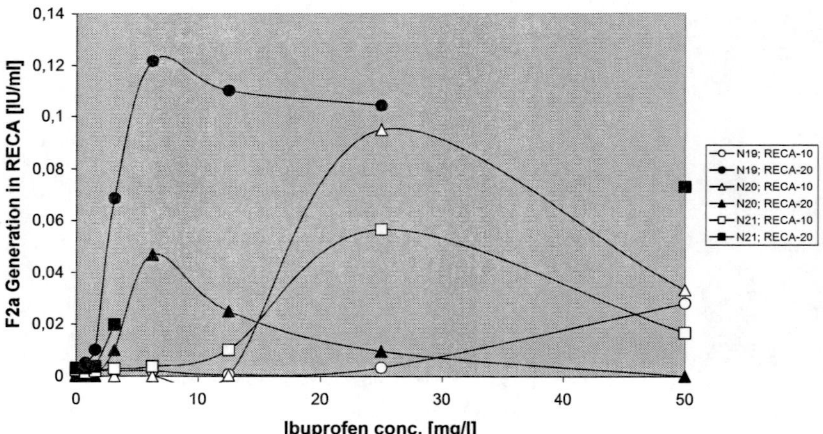

Figure 16. Thrombin (F2a) generation by ibuprofen in individual plasmas 19-21. 0-50 mg/l ibuprofen (sodium salt) were added to 50 µl individual platelet poor citrated plasma (frozen/thawed) in high quality polystyrene U-wells (Brand®781600). The recalcified coagulation activity assay (RECA) was performed with 10 min (white symbols) or 20 min (black symbols) coagulation reaction time (RECA-10 or RECA-20). The approx. SC200 values were 1, 2, 2 mg/l ibuprofen.

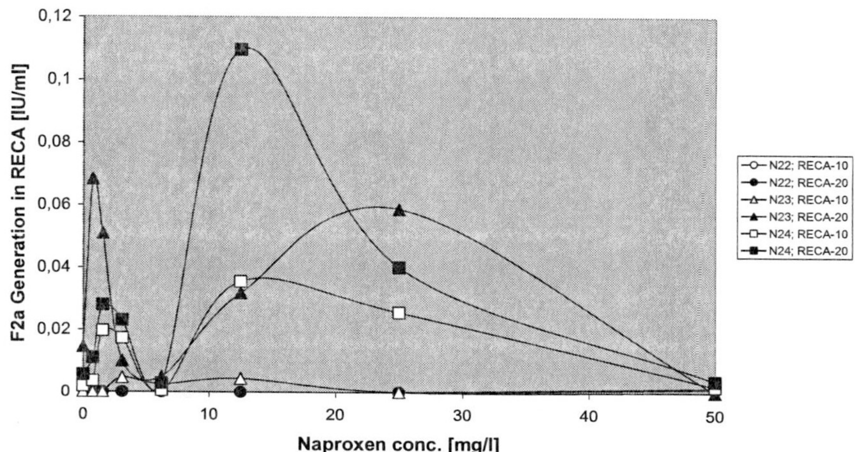

Figure 17. Thrombin (F2a) generation by naproxen in individual plasmas 22-24. 0-50 mg/l naproxen (sodium salt) were added to 50 µl individual platelet poor citrated plasma (frozen/thawed) in high quality polystyrene U-wells (Brand®781600). The recalcified coagulation activity assay (RECA) was performed with 10 min (white symbols) or 20 min (black symbols) coagulation reaction time (RECA-10 or RECA-20). The approx. SC200 values were 70, 0.4, 0.8 mg/l naproxen.

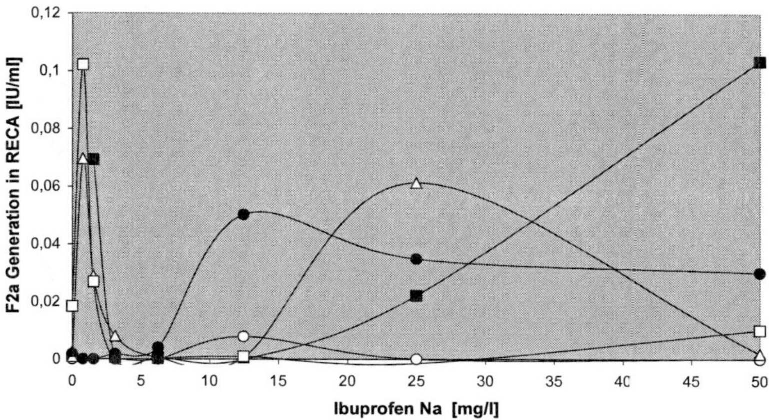

Figure 18. Thrombin (F2a) generation by ibuprofen in individual plasmas 22-24. 0-50 mg/l ibuprofen (sodium salt) were added to 50 µl individual platelet poor citrated plasma (frozen/thawed) in high quality polystyrene U-wells (Brand®781600). The recalcified coagulation activity assay (RECA) was performed with 10 min (white symbols) or 20 min (black symbols) coagulation reaction time (CRT; RECA-10 or RECA-20). The approx. SC200 values were 5, 0.1, 0.2 mg/l ibuprofen.

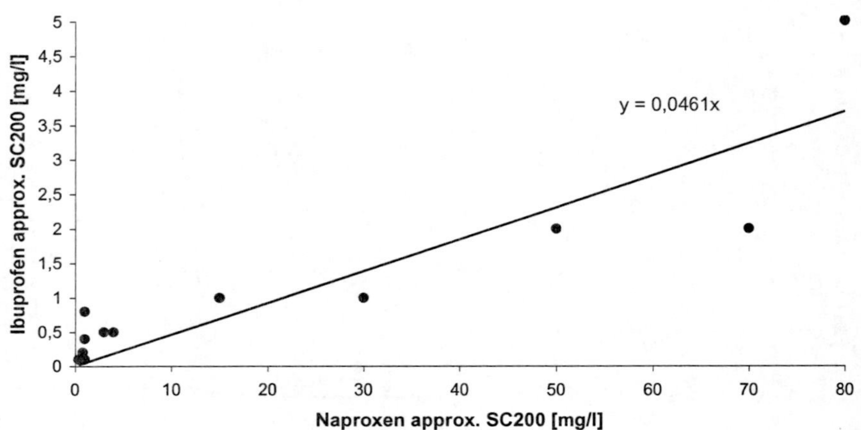

Figure 19. Comparison of the approx. SC200 of naproxen with that of ibuprofen. 12 individual plasmas were supplemented with 0-50 mg/l naproxen or ibuprofen as indicated in figures 11-18. The approx. SC200 values on recalcified thrombin generation were compared: 21±29 mg/l (MV±1SD) naproxen vs. 1.1±1.4 mg/l ibuprofen; r= 0.897.

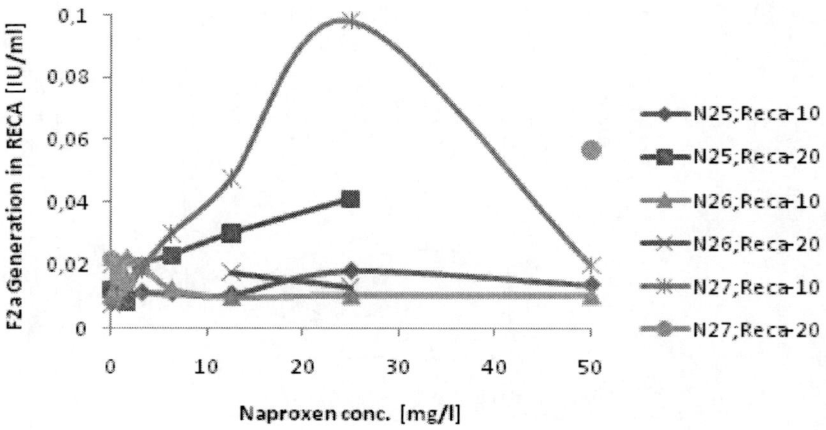

Figure 20. Thrombin (F2a) generation by naproxen in individual plasmas 25-27. 0-50 mg/l naproxen (sodium salt) were added to 50 µl individual platelet poor citrated plasma (frozen/thawed) in high quality polystyrene U-wells (Brand®781600). The recalcified coagulation activity assay (RECA) was performed with 10 min or 20 min coagulation reaction time (RECA-10 or RECA-20). The approx. SC200 values were 4, 0.8, 2 mg/l naproxen.

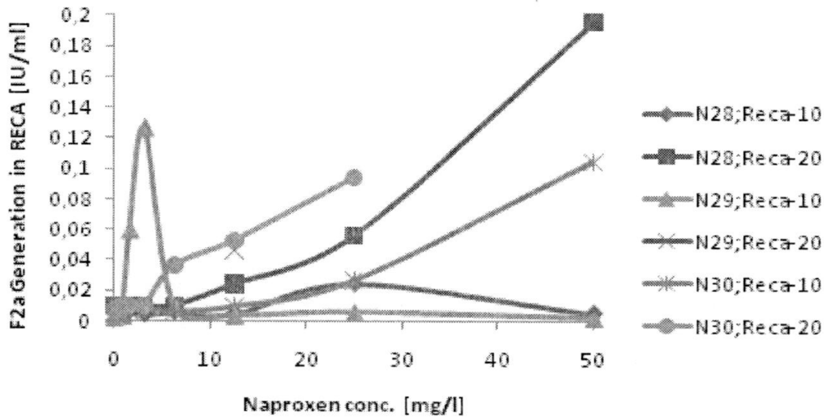

Figure 21. Thrombin (F2a) generation by naproxen in individual plasmas 28-30. 0-50 mg/l naproxen (sodium salt) were added to 50 µl individual platelet poor citrated plasma (frozen/thawed) in high quality polystyrene U-wells (Brand®781600). The recalcified coagulation activity assay (RECA) was performed with 10 min or 20 min coagulation reaction time (RECA-10 or RECA-20). The approx. SC200 values were 10, 1, 10 mg/l naproxen.

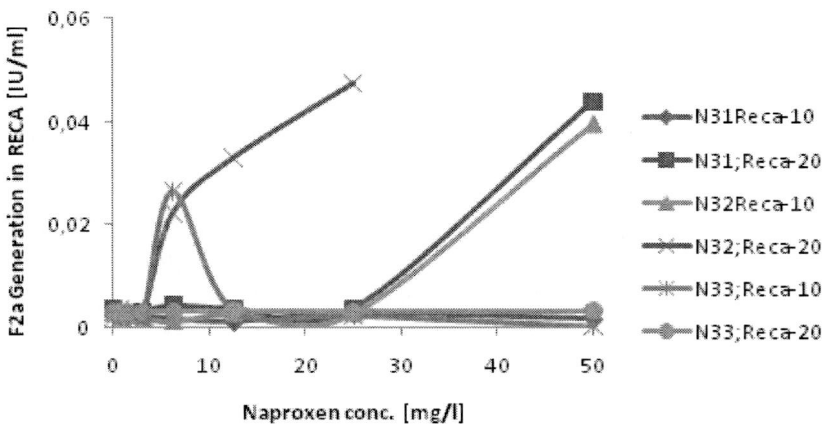

Figure 22. Thrombin (F2a) generation by naproxen in individual plasmas 31-33. 0-50 mg/l naproxen (sodium salt) were added to 50 µl individual platelet poor citrated plasma (frozen/thawed) in high quality polystyrene U-wells (Brand®781600). The recalcified coagulation activity assay (RECA) was performed with 10 min or 20 min coagulation reaction time (RECA-10 or RECA-20). The approx. SC200 values were 30, 4, 80 mg/l naproxen.

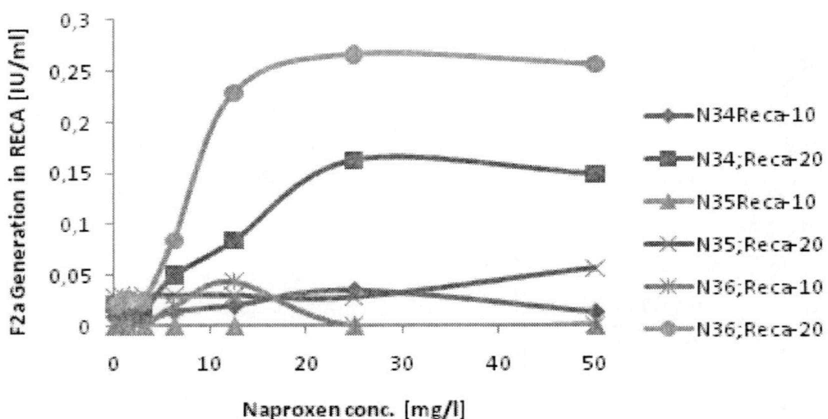

Figure 23. Thrombin (F2a) generation by naproxen in individual plasmas 34-36. 0-50 mg/l naproxen (sodium salt) were added to 50 µl individual platelet poor citrated plasma (frozen/thawed) in high quality polystyrene U-wells (Brand®781600). The recalcified coagulation activity assay (RECA) was performed with 10 min or 20 min coagulation reaction time (RECA-10 or RECA-20). The approx. SC200 values were 5, 50, 4 mg/l naproxen.

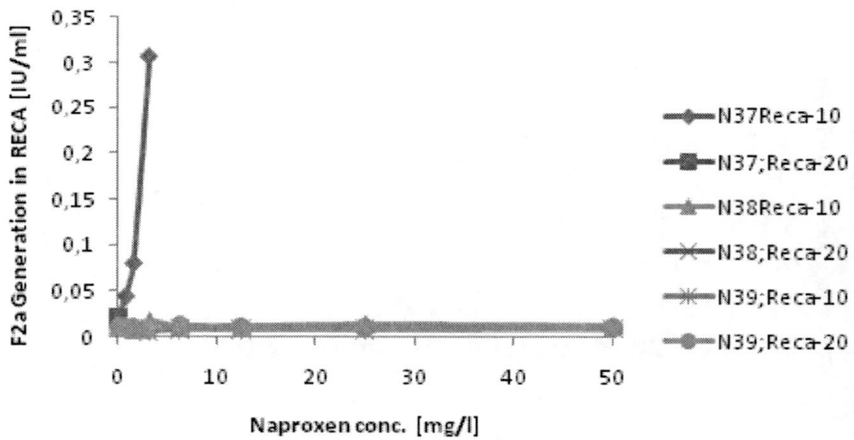

Figure 24. Thrombin (F2a) generation by naproxen in individual plasmas 37-39. 0-50 mg/l naproxen (sodium salt) were added to 50 µl individual platelet poor citrated plasma (frozen/thawed) in high quality polystyrene U-wells (Brand®781600). The recalcified coagulation activity assay (RECA) was performed with 10 min or 20 min coagulation reaction time (RECA-10 or RECA-20). The approx. SC200 values were 0.6, 80, 90 mg/l naproxen.

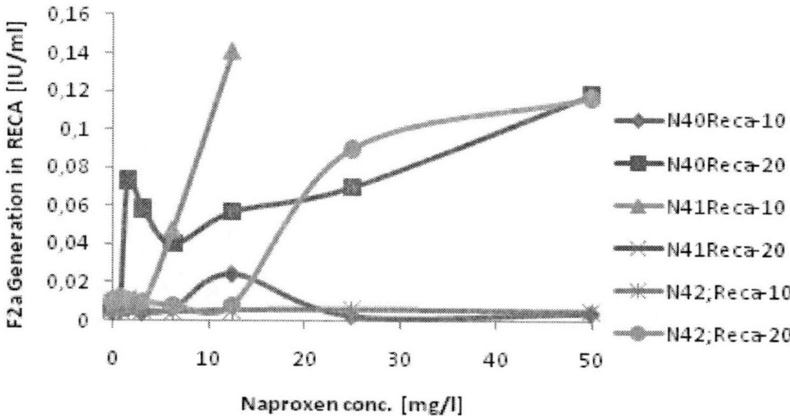

Figure 25. Thrombin (F2a) generation by naproxen in individual plasmas 40-42. 0-50 mg/l naproxen (sodium salt) were added to 50 µl individual platelet poor citrated plasma (frozen/thawed) in high quality polystyrene U-wells (Brand®781600). The recalcified coagulation activity assay (RECA) was performed with 10 min or 20 min coagulation reaction time (RECA-10 or RECA-20). The approx. SC200 values were 1, 4, 15 mg/l naproxen.

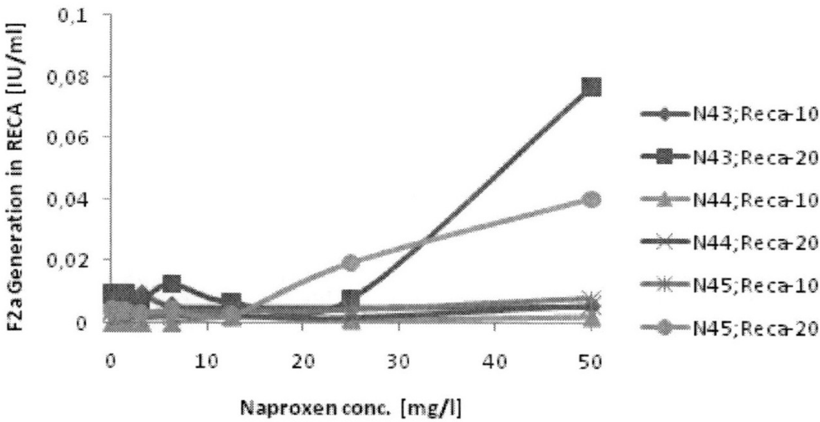

Figure 26. Thrombin (F2a) generation by naproxen in individual plasmas 43-45. 0-50 mg/l naproxen (sodium salt) were added to 50 µl individual platelet poor citrated plasma (frozen/thawed) in high quality polystyrene U-wells (Brand®781600). The recalcified coagulation activity assay (RECA) was performed with 10 min or 20 min coagulation reaction time (RECA-10 or RECA-20). The approx. SC200 values were 30, 65, 20 mg/l naproxen.

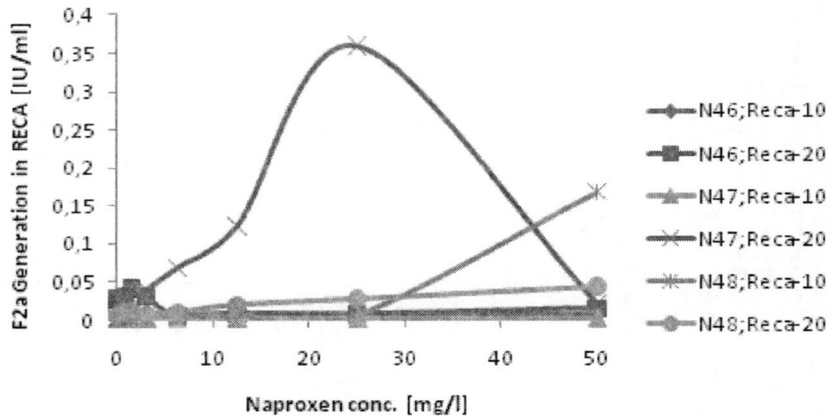

Figure 27. Thrombin (F2a) generation by naproxen in individual plasmas 46-48. 0-50 mg/l naproxen (sodium salt) were added to 50 µl individual platelet poor citrated plasma (frozen/thawed) in high quality polystyrene U-wells (Brand®781600). The recalcified coagulation activity assay (RECA) was performed with 10 min or 20 min coagulation reaction time (RECA-10 or RECA-20). The approx. SC200 values were 3, 1, 2 mg/l naproxen.

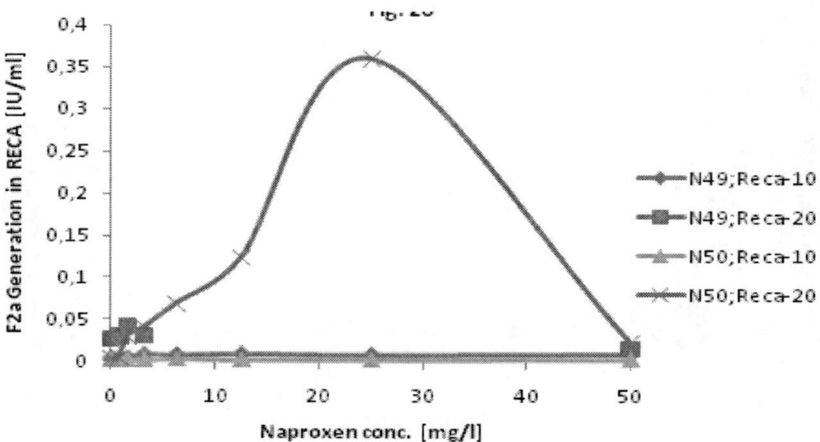

Figure 28. F2a generation by naproxen in individual plasmas 49 and 50. 0-50 mg/l naproxen (sodium salt) were added to 50 µl individual platelet poor citrated plasma (frozen/thawed) in high quality polystyrene U-wells (Brand®781600). The recalcified coagulation activity assay (RECA) was performed with 10 min or 20 min coagulation reaction time (RECA-10 or RECA-20). The approx. SC200 values were 3 and 1 mg/l naproxen.

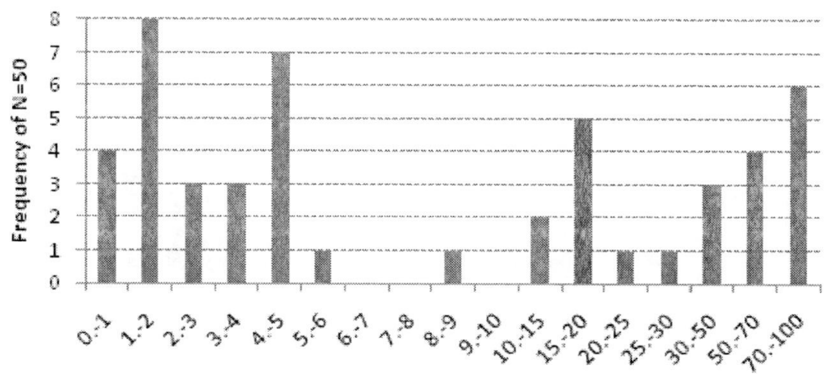

Figure 29. Distribution of naproxen approx. SC200 among 50 normal plasmas. 24% of the plasmas were susceptible to naproxen induced AM-coagulation (approx. SC200 < 2 mg/l naproxen; 0.4 mg/l = lowest approx. SC200), 30% of the plasmas behaved as intermediate sensible (approx. SC200 of 2-8 mg/l naproxen), and 46% of plasmas were rather resistant against naproxen induced AM-coagulation (approx. SC200 ≥ 8 mg/l naproxen). For all samples the approx. SC200 was 20±27 mg/l naproxen (MV±1SD).

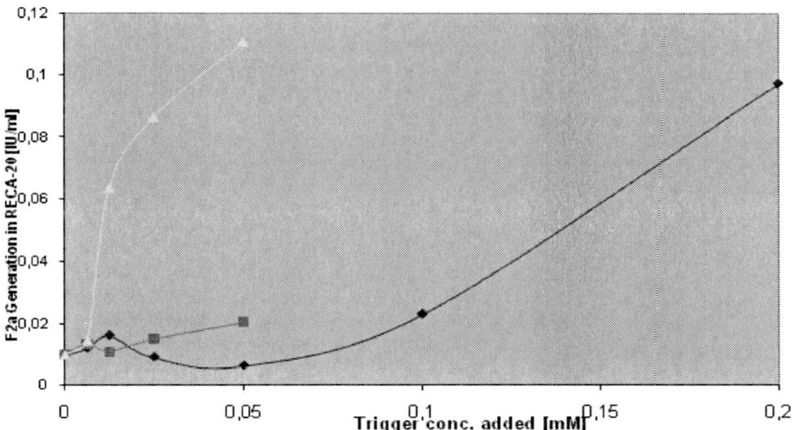

Figure 30. Thrombin (F2a) generation in dependence of ketone bodies in normal plasma 1. 50 µl normal citrated plasma (about 3h 23°C old) were supplemented in U-microwells with 2 µl 0-5.2 mM 3-HB (♦), acetoacetate (■), or acetone (▲). Then the RECA was performed with a coagulation reaction time of 20 min (RECA-20). The approx. SC200 were 0.1 mM 3-HB, 0.05 mM acetoacetate, or 0.01 mM acetone.

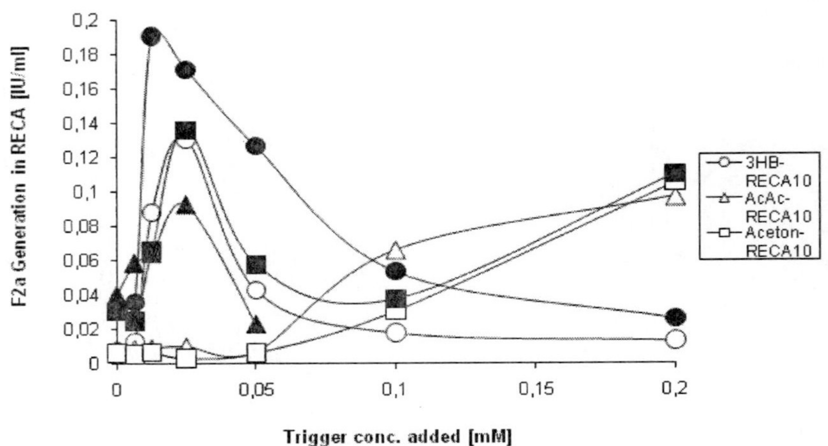

Figure 31. Thrombin (F2a) generation in dependence of ketone bodies in normal plasma 2. 50 µl normal human citrated plasma (about 3h 23°C old) were supplemented in U-microwells (Brand®781600) with 2 µl 0-5.2 mM 3-HB (O●), acetoacetate (Δ▲), or acetone (□■). Then the RECA-10 or RECA-20 was performed. The approx. SC200 were 0.01 mM 3-HB, 0.02 mM acetoacetate, or 0.01 mM acetone.

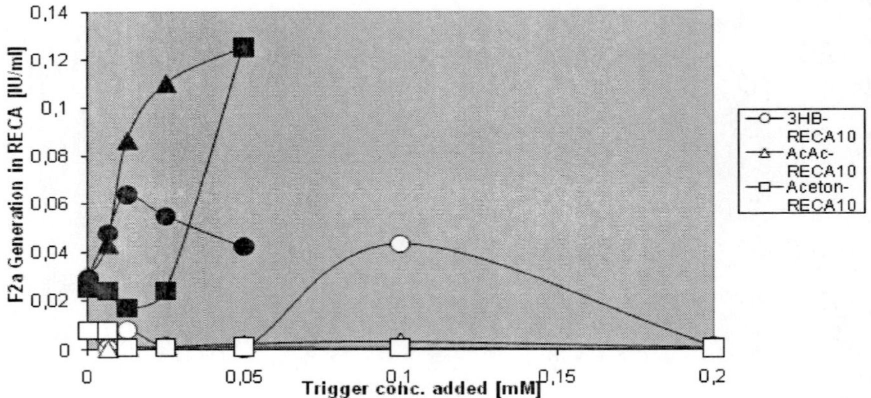

Figure 32. Thrombin (F2a) generation in dependence of ketone bodies in normal plasma 3. 50 µl normal human citrated plasma (about 3h 23°C old) were supplemented in U-microwells (Brand®781600) with 2 µl 0-5.2 mM 3-hydroxy-butyrate (3-HB;O●), acetoacetate (Δ▲), or acetone (□■). Then the RECA-10 or RECA-20 was performed. The approx. SC200 were 0.01 mM 3-HB, 0.01 mM acetoacetate, or 0.03 mM acetone.

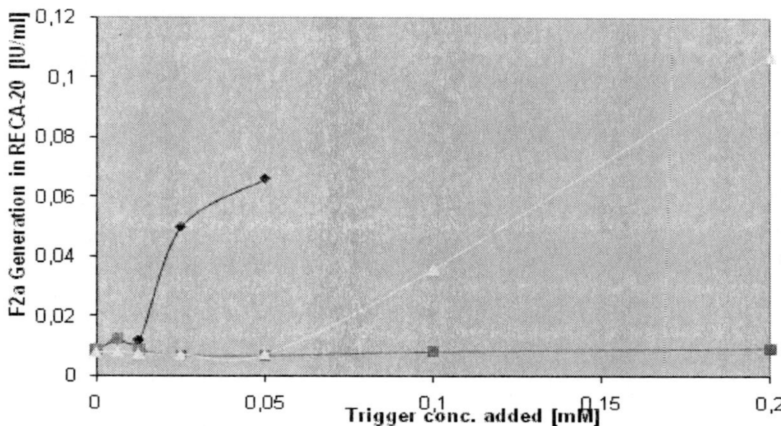

Figure 33. Thrombin (F2a) generation in dependence of ketone bodies in normal plasma 4. 50 µl normal human citrated plasma (about 3h 23°C old) were supplemented in U-microwells (Brand®781600) with 2 µl 0-5.2 mM 3-hydroxy-butyrate (3-HB;●), acetoacetate (■), or acetone (▲). Then the RECA was performed with a coagulation reaction time of 20 min (RECA-20). The approx. SC200 were 0.015 mM 3-HB, > 0.2 mM acetoacetate, or 0.07 mM acetone.

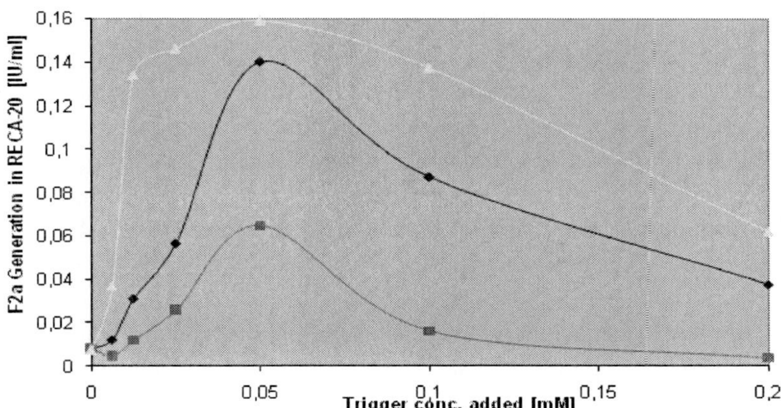

Figure 34. Thrombin (F2a) generation in dependence of ketone bodies in normal plasma 5. 50 µl normal human citrated plasma (about 3h 23°C old) were supplemented in U-microwells (Brand®781600) with 2 µl 0-5.2 mM 3-hydroxy-butyrate (3-HB;●), acetoacetate (■), or acetone (▲). Then the RECA was performed with a coagulation reaction time of 20 min (RECA-20). The approx. SC200 were 0.01 mM 3-HB, 0.02 mM acetoacetate, or 0.005 mM acetone.

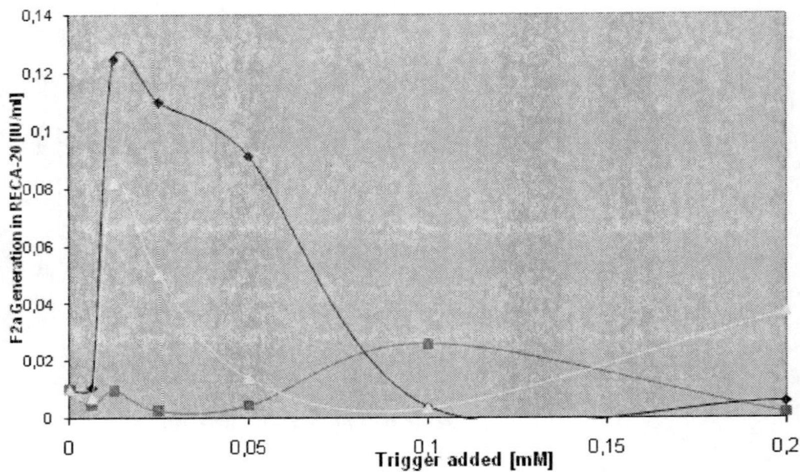

Figure 35. Thrombin (F2a) generation in dependence of ketone bodies in normal plasma 6. 50 µl normal human citrated plasma (about 3h 23°C old) were supplemented in U-microwells (Brand®781600) with 2 µl 0-5.2 mM 3-hydroxy-butyrate (3-HB;●), acetoacetate (■), or acetone (▲). Then the RECA was performed with a coagulation reaction time of 20 min (RECA-20). The approx. SC200 were 0.01 mM 3-HB, 0.08 mM acetoacetate, or 0.01 mM acetone.

As typical for contact activation [33-36], there is great inter-individual variation of the susceptibility towards the AM-trigger naproxen. Patients on naproxene with a low naproxen SC200 on intrinsic coagulation and with additional thrombosis risks such as insufficient blood flow or hepatic insufficiency [37] need low-molecular-weight-heparin (LMWH) prophylaxis to compensate for the increased thrombin generation risk and to stimulate cellular fibrinolysis (just in case of pathologic micro-thrombi) [38-42].

There are 3 types of prostaglandins [10]:

- Series-1-prostaglandins (derived of dihomo-γ-linolenic acid (DGLA = 20:3 (Figure 2): usually anti-inflammatory and anti-platelet activating.
- Series-2-prostaglandins (derived of arachidonic acid = AA =20:4 (Figure 3): usually pro-inflammatory.
- Series-3-prostaglandins (derived of eicosapentaenoic acid (EPA =20:5 (Figure 4)). DGLA/EPA compete with AA for COX.

The prostaglandin receptors are G-Protein-coupled 7-transmembrane rezeptors (similar to opsin). They are named P (for prostaglandin receptor) with D, E, F, I, T specificity for the prostaglandins D, E, F, I, thromboxane.

Furthermore, there are prostaglandins (e.g. PGJ_2), that activate nucleus receptors, that inhibit the NF-κB pathway (via inhibition of IκB-kinase) [10].

Prostaglandins are of enormous importance in different biological systems. Additional work on naproxen and blood ROS in refined test systems will clarify the action of naproxen on neutrophils [43-47].

Any drug can trigger (contact phase) altered matrix (AM) - coagulation in susceptible patients, resulting in a critical NIC→PIC-0 change (normal intravascular coagulation enters the pre-phase of pathologic intravascular coagulation) [48]. Consequently, pathological concentrations of systemically circulating micro-thrombi are generated that can be deposited in any organ. The ischemia would cause cell necrosis in the heart (with TNI values of about 0.1 μg/l; N< 0.04 μg/l), in the kidneys (with two-fold elevated creatinine conc.), in the liver (with two-fold elevated conc. of total bilirubin or GGT), in the pancreas (with slightly increased lipase or amylase conc.), in the lung (with slightly pathologically changed blood gases) or in the brain (with changes in general vigilance). Susceptible patients with systemic activation of AM-coagulation and pathologic micro-thrombi should never be treated by modern oral thrombin/F10a inhibitors [49,50]. Here the physiologic LMWH is the preferred drug for thrombosis prophylaxis and therapy.

Acknowledgments

The medical students of the Clinical Biochemistry course 2012/2013 of the University Hospital of Marburg are thanked for giving blood samples after written informed consent. There are no conflicts of interest and there was no specific funding.

References

[1] Burmester G, Lanas A, Biasucci L, Hermann M, Lohmander S, Olivieri I, Scarpignato C, Smolen J, Hawkey C, Bajkowski A, Berenbaum F, Breedveld F, Dieleman P, Dougados M, MacDonald T, Mola EM, Mets T, Van den Noortgate N, Stoevelaar H. The appropriate use of non-steroidal anti-inflammatory drugs in rheumatic disease: opinions of a multidisciplinary European expert panel. *Ann Rheum Dis*. 2011; 70: 818-22.

[2] Lai LH, Chan FK. Nonsteroid anti-inflammatory drug-induced gastroduodenal injury. *Curr Opin Gastroenterol*. 2009; 25: 544-8.

[3] Derry C, Derry S, Moore RA, McQuay HJ. Single dose oral naproxen and naproxen sodium for acute postoperative pain in adults. *Cochrane Database Syst Rev*. 2009; (1):CD004234.

[4] Cleves C, Tepper SJ. Sumatriptan/naproxen sodium combination for the treatment of migraine. *Expert Rev Neurother*. 2008; 8: 1289-97.

[5] Pöllmann W, Feneberg W. Current management of pain associated with multiple sclerosis. *CNS Drugs*. 2008; 22: 291-324.

[6] Salvi GE, Lang NP. The effects of non-steroidal anti-inflammatory drugs (selective and non-selective) on the treatment of periodontal diseases. *Curr Pharm Des*. 2005; 11: 1757-69.

[7] McGettigan P, Henry D. Cardiovascular risk with non-steroidal anti-inflammatory drugs: systematic review of population-based controlled observational studies. *PLoS Med*. 2011; 8: e1001098.

[8] Varas-Lorenzo C, Riera-Guardia N, Calingaert B, Castellsague J, Pariente A, Scotti L, Sturkenboom M, Perez-Gutthann S. Stroke risk and NSAIDs: a systematic review of observational studies. *Pharmacoepidemiol Drug Saf*. 2011; 20: 1225-36.

[9] Tegeder I, Geisslinger G. Cardiovascular risk with cyclooxygenase inhibitors: general problem with substance specific differences? *Naunyn Schmiedebergs Arch Pharmacol*. 2006; 373:1-17.

[10] www.wikipedia.org

[11] Agúndez JA, García-Martín E, Martínez C. Genetically based impairment in CYP2C8- and CYP2C9-dependent NSAID metabolism as a risk factor for gastrointestinal bleeding: is a combination of pharmacogenomics and metabolomics required to improve personalized medicine? *Expert Opin Drug Metab Toxicol*. 2009; 5: 607-20.

[12] Rodrigues AD. Impact of CYP2C9 genotype on pharmacokinetics: are all cyclooxygenase inhibitors the same? *Drug Metab Dispos*. 2005; 33: 1567-75.

[13] van Leeuwen JS, Unlü B, Vermeulen NP, Vos JC. Differential involvement of mitochondrial dysfunction, cytochrome P450 activity, and active transport in the toxicity of structurally related NSAIDs. *Toxicol In Vitro*. 2012; 26: 197-205.

[14] Ali S, Pimentel JD, Ma C. Naproxen-induced liver injury. *Hepatobiliary Pancreat Dis Int*. 2011; 10: 552-6.

[15] Trelle S, Reichenbach S, Wandel S, Hildebrand P, Tschannen B, Villiger PM, Egger M, Jüni P. Cardiovascular safety of non-steroidal anti-inflammatory drugs: network meta-analysis. *BMJ* 2011; 342: c7086.

[16] Stief TW. Neutrophil granulocytes in hemostasis. *Hemostasis Laboratory* 2008; 1: 269-89.

[17] Mena J, Manosalva C, Ramirez R, Chandia L, Carroza D, Loaiza A, Burgos RA, Hidalgo MA. Linoleic acid increases adhesion, chemotaxis, granule release, intracellular calcium mobilisation, MAPK phosphorylation and gene expression in bovine neutrophils. *Vet Immunol Immunopathol.* 2013; 151: 275-84.

[18] Kantarci A, Oyaizu K, Van Dyke TE. Neutrophil-mediated tissue injury in periodontal disease pathogenesis: findings from localized aggressive periodontitis. *J Periodontol.* 2003; 74: 66-75.

[19] Bennett A, Villa G. Nimesulide: an NSAID that preferentially inhibits COX-2, and has various unique pharmacological activities. *Expert Opin Pharmacother.* 2000; 1: 277-86.

[20] Stief TW. Drug - induced thrombin generation : the breakthrough. *Hemostasis Laboratory* 2010; 3: 3-6.

[21] Stief TW. Pathological thrombin generation by the synthetic inhibitor argatroban. *Hemost Lab.* 2009; 2: 83-104.

[22] Stief T. Singlet oxygen ($^1O_2^*$) primes blood neutrophils to generate ROS. *Hemostasis Laboratory* 2013; 6 (issue 4)

[23] Graff J, Skarke C, Klinkhardt U, Watzer B, Harder S, Seyberth H, Geisslinger G, Nüsing RM. Effects of selective COX-2 inhibition on prostanoids and platelet physiology in young healthy volunteers. *J Thromb Haemost.* 2007; 5: 2376-85.

[24] Mosesson MW. Update on antithrombin I (fibrin). *Thromb Haemost.* 2007; 98: 105-8.

[25] Stief TW. The recalcified coagulation activity. *Clin Appl Thrombosis Hemostasis.* 2008; 14: 447-53.

[26] Stief TW. Thrombin triggers its generation in individual platelet poor plasma. *Hemost Lab.* 2009; 2: 363-78.

[27] Stief TW. Thrombin generation by diclofenac. In: *Thrombin: function and pathophysiology*. Stief T, ed.; NOVA science publishers; New York; 2012; pp. 73-83.

[28] Stief TW. Thrombin generation by ibuprofen. *Hemostasis Laboratory* 2010; 3: 277-82.

[29] Stief TW. Contact activation of coagulation depends on the maximal lipophilic trigger concentration. *Blood Coagulation and Fibrinolysis* 2012; 23: 296-8.

[30] Stief TW. The maximal plasma concentration of (delta-) negatively charged contact triggers influences plasmatic thrombin generation.

In: *Thrombin: function and pathophysiology*. Stief T, ed.; NOVA science publishers; New York; 2012; pp. 37-46.

[31] Stief TW, Brödje D. Thrombin generation by 3-hydroxy-butyrate, acetoacetate, or acetone. *Hemostasis Laboratory* 2013; 6: 17-50.

[32] Taha MO, Dahabiyeh LA, Bustanji Y, Zalloum H, Saleh S. Combining ligand-based pharmacophore modeling, quantitative structure-activity relationship analysis and in silico screening for the discovery of new potent hormone sensitive lipase inhibitors. *J Med Chem*. 2008; 51: 6478–94.

[33] Stief TW. Thrombin generation by therapeutic fibrinogen. *Hemost Lab*. 2011; 4: 467-82.

[34] Stief TW. Thrombin generation triggered by ultra-low-concentrations of heparins. *Hemost Lab*. 2011; 4: 65-76.

[35] Stief TW. Xenobiotic - induced pancreas carcinoma following mesenteric vein thrombosis : a hypothesis. *Hemost Lab*. 2010; 3: 253-8.

[36] Stief TW. Thrombin generation by creatinine. *Hemostasis Laboratory* 2011; 4: 191-9.

[37] Loureiro-Silva MR, Kouyoumdjian M, Borges DR. Plasma kallikrein and thrombin are cleared through unrelated hepatic pathways. *Blood Coagul Fibrinolysis* 1993; 4: 551-4.

[38] Stief TW. The efficiency of anti-factor Xa and anti-factor IIa anticoagulants. *Blood Coagul Fibrinolysis* 2007; 18: 265-9.

[39] Stief TW. Thrombin – applied clinical biochemistry of the main factor of coagulation. In: *Thrombin: function and pathophysiology*. Stief T, ed.; Nova science publishers; New York; 2012; pp. vii-xx.

[40] Stief TW. LMWH - action-monitoring for all patients.*Acta Paediatr*. 2012; 101: e314. 10.1111/j.1651-2227.2012.02712.x.

[41] Stief TW. Kallikrein activates prothrombin. *Clin. Appl. Thromb. Hemost*. 2008; 1: 97-8.

[42] Stief TW. Kallikrein activates factor 10. *Hemost. Lab*. 2012; 5: 211-8.

[43] Costa D, Gomes A, Lima JL, Fernandes E. Singlet oxygen scavenging activity of non-steroidal anti-inflammatory drugs. *Redox Rep*. 2008; 13: 153-60.

[44] Costa D, Moutinho L, Lima JL, Fernandes E. Antioxidant activity and inhibition of human neutrophil oxidative burst mediated by arylpropionic acid non-steroidal anti-inflammatory drugs.*Biol Pharm Bull*. 2006; 29: 1659-70.

[45] Paino IM, Ximenes VF, Fonseca LM, Kanegae MP, Khalil NM, Brunetti IL. Effect of therapeutic plasma concentrations of non-steroidal anti-

inflammatory drugs on the production of reactive oxygen species by activated rat neutrophils. *Braz J Med Biol Res*. 2005; 38: 543-51.

[46] Parij N, Nagy AM, Fondu P, Nève J. Effects of non-steroidal anti-inflammatory drugs on the luminol and lucigenin amplified chemiluminescence of human neutrophils. *Eur J Pharmacol.* 1998; 352: 299-305.

[47] Loike JD, Silverstein R, Wright SD, Weitz JI, Huang AJ, Silverstein SC. The role of protected extracellular compartments in interactions between leukocytes, and platelets, and fibrin/fibrinogen matrices. *Ann NY Acad Sci* 1992; 667: 163-72.

[48] Stief TW. The laboratory diagnosis of the pre-phase of pathologic disseminated intravascular coagulation. *Hemostasis Laboratory* 2008; 1: 2-20.

[49] http://www.docguide.com_2012_12_21. FDA warns against use of dabigatran in patients with mechanical heart valves.

[50] Stief TW. Rivaroxaban or dabigatran at less than 20% of their therapeutic plasma concentration can pathologically trigger thrombin generation. *Hemostasis Laboratory* 2013; 6: 65-76.

INDEX

A

B

C

D

E

F

G

H

I

J

K

L

M

N

O

P

R

S

T

U

V

W